THE
BMW
STORY

THE
BMW
STORY

Haynes

PRODUCTION AND RACING MOTORCYCLES FROM 1923 TO THE PRESENT DAY
IAN FALLOON

First published in November 2003

A catalogue record for this book is available from the British Library

ISBN 1 85960 854 X

Library of Congress catalog card no. 2003110427

Published by Haynes Publishing, Sparkford,
Yeovil, Somerset, BA22 7JJ, UK

Tel: 01963 442030 Fax: 01963 440001
Int. tel: +44 1963 442030 Int. fax: +44 1963 440001
E-mail: sales@haynes.co.uk
Web site: www.haynes.co.uk

Haynes North America, Inc.,
861 Lawrence Drive, Newbury Park,
California 91320, USA

Printed and bound in England by J. H. Haynes & Co. Ltd, Sparkford

CONTENTS

Following on from the R90S and R100S, the R100CS incorporated the attributes of the early S models, with the technical improvements of the post–1981 boxers. *Ian Falloon*

INTRODUCTION & ACKNOWLEDGEMENTS

Over the past 80 years, BMW motorcycles have provided a unique alternative to those of other manufacturers. Some motorcycles may have been faster, certainly others were cheaper, but with their emphasis on quality and reliability, none has emulated the practicality of a BMW. With its commitment to ease of serviceability, a BMW has become the preferred choice for hundreds of thousands of motorcyclists around the world.

Although more recently built in Berlin, the BMW motorcycle tradition began in Munich, the capital of Bavaria in Southern Germany. Consequently, the design of BMW motorcycles was well suited to the twisting roads of the Bavarian countryside and the nearby Alps. When Schorsch Meier, winner of the 1939 Senior TT on a 500 Kompressor, was asked how he could lap the Isle of Man at more than 90mph (145km/h) with virtually no track experience, he replied that it reminded him of the roads around his home in Bavaria. When production recommenced after the Second World War, BMW continued this tradition of building motorcycles that excelled on the back roads, in the Alps, or on the Autobahn, a tradition which continues through to today. With the establishment of the dual-purpose GS (Gelände/Straße) the parameters widened to include off-road use.

BMWs are to be ridden, and if the journey includes a mixture of straight roads and corners, there is no better motorcycle. Take into account the best range of factory luggage and accessories available, and the success of BMW motorcycles isn't surprising. Ownership of a BMW is also generally a long-term affair. Because of long model runs, the designs are so well developed there seems little need to update on the whim of fashion. This is most definitely a refreshing change from the current trend of instant obsolescence. My own experience with BMW motorcycles extends from many wonderful long rides on a friend's early model R100RS, to long-term ownership of a K100RS. Advancing age has seen a reversion to the traditional boxer twin, and my 1976 R90S continues to provide enjoyment. Over the years, I have had the fortune to sample a wide range of BMW motorcycles, although I must admit to finding the huge size of some of the newer examples intimidating.

While I have endeavoured to document every road and racing BMW motorcycle since 1923, I have also tried to provide an insight into the people responsible for the designs. Without Max Friz, Rudolf Schleicher, Alfred

Böning, Alexander von Falkenhausen, Hans-Günther von der Marwitz, Stefan Pachernegg, and David Robb, BMW motorcycles wouldn't be what they are.

Compiling this book would have been impossible without the help of many friends, colleagues, and enthusiasts. I have endeavoured to find as much previously unpublished photographic material as possible, and for this I am indebted to Fred Jakobs, Ruth Standfuss, and Annette Schultz of BMW Mobile Tradition in Munich. Brian J. Nelson has contributed wonderful images of older BMW motorcycles from collections in the USA, and Mick Woollett has allowed the generous use of his comprehensive archive.

Although there have been several books already published on the history of BMW motorcycles, many contradictions led me to consult the absolute expert on BMW motorcycle history: Stefan Knittel. Stefan spent two years in the BMW archives researching his milestone book *BMW Motorräder* (1984), and his knowledge is unsurpassed. He agreed to read the manuscript, and supply several photographs, so hopefully, this book will present the most accurate history available in English.

I would also like to thank Ken Wootton, the editor of *Australian Motorcycle News*, who not only allowed the use of photos from their archives, but also many years of BMW press releases. David Edwards, editor of *Cycle World* magazine, generously provided some important archival photographs. Damien Cook, of the BMWMCC Victoria, was tireless in his support and enthusiasm for this project, organising a wide range of motorcycles to be available for photography. Further thanks go to Bruce Armstrong, Julian Barson, Peter Cullen, Richard Fanning, Lloyd Griffiths, Ian Isbister, Chris McArdle, Metzger Rudl, Rainer Schwinge, Darryl Suter, Heinz Tschinkel, and Ken Wright. Everyone was so cooperative and enthusiastic it was a humbling experience, and if anyone has been omitted, I sincerely apologise.

I would also like to thank Mark Hughes's excellent and professional editorial team at Haynes – they are a pleasure to work with. Finally, the tireless hours spent researching and writing this book would not have been possible without the wonderful support from my wife Miriam, and two sons, Benjamin and Timothy.

Ian Falloon
July 2003

The R32 of 1923. With its boxer
engine and shaft drive, it
established a layout that would
epitomise most BMW motorcycles
for the next 80 years. *Cycle World*

1 BEGINNINGS
FROM AIRCRAFT TO MOTORCYCLES

In the company's register in the Munich District Court, there is an entry in copperplate handwriting for 7 March 1916, listing the Bavarian Aircraft Works, a company to manufacture and market aircraft. One of the six investors listed was Gustav Otto, an engineer, pilot, flying instructor, and entrepreneur. Gustav came from an illustrious background and his father, Nikolaus August Otto, was the creator of the four-stroke petrol engine. When Gustav was only eight years old, in 1891, Nikolaus died, but he would follow his father's footsteps into engineering and when he saw Louis Blèriot flying in 1909, he was captivated.

In 1910, Gustav formed his own aircraft manufacturing works in Munich and by 1912 he was one of Germany's most prominent air pioneers. Unfortunately, just like his father, Gustav was a poor businessman and the only financial partnership he could arrange was with the German military. Even war couldn't save Gustav and during 1915 his relationship with the German War Ministry disintegrated following a disagreement with one of Germany's leading generals, von Brug. Forced into bankruptcy, Gustav spent some time in a mental hospital, then he sold the Otto Works and the premises in Oberwiesenfeld. These premises, at 76 Neulerchenfeld Street, later renamed Lerchenauer Street, are still the site for Plant 1 at BMW AG. Although he then severed all ties with the new company, ultimately shooting himself in 1926, Gustav Otto is credited as being BMW's founding father. Yet, this is only where half of the BMW story began.

The other half commenced in another corner of Oberwiesenfeld, in 1912, when Karl Rapp established a Munich subsidiary of the Aachen-based Flugwerke Deutschland to manufacture aircraft engines. Unlike Otto, Rapp couldn't boast an illustrious ancestry, but he came at a time when Germany suddenly discovered they were falling behind France in the development of military aviation. Kaiser Wilhelm II established an Imperial Prize for an aviation engine, but Rapp's entry failed. The firm went into receivership and Rapp managed to raise 200,000 marks to establish the Rapp Motor Works in the same premises at 288 Schleiflheim Street, on 28 October 1913.

With six workers, Rapp again prepared an engine for the Imperial Prize. This time, it was a six-cylinder, but, plagued with excessive vibration, the engine was not completed on time. The outbreak of war gave

Max Friz, the father of the BMW
boxer motorcycle. *Stefan Knittel*

MAX FRIZ

Born in Urach in 1883, Max Friz was apprenticed at the age
of 15 to the Kuhn steam engine company in Cannstatt. In
1902, he went to the Royal Building Trade School in
Stuttgart-Esslingen and from 1906 worked in the design
office of the Daimler-Motoren-Gesellschaft. Here, he made a
significant contribution to the design of the racing engine
for the 1914 Mercedes Grand Prix car that won the French
Grand Prix. Between 1912 and 1913, Friz designed the first
German aircraft engine with separate cylinders on the
crankcase, and an overhead camshaft driven by a vertical
shaft and bevel gears. Following the release of the R32
motorcycle, Friz became a director of BMW AG, and from
1925 until 1937 was their first chief designer. He was general
manager of the Munich plant from 1935 until 1937, and then
held a similar position at Eisenach until 1944. Friz retired in
1945 and died in 1966.

Rapp a reprieve and he soon produced V8 and V12
aero engines. Unfortunately, these also vibrated and
they were rejected by the Prussian military command in
Berlin. Like the Otto Works down the road, Rapp's
company also suffered from poor business management
and his new commercial director, Max Wiedmann,
steered the company towards the brink of disaster. The
workforce grew to 370, but by the autumn of 1916,
the Austrian military ceased to be a customer because
of the poor reputation of Rapp's engines. Wiedmann
befriended a Vienna-based financier (and notorious
war profiteer) Camillo Castiglioni who already
controlled Austro-Daimler in Austria. Austro-Daimler
didn't have the capacity to supply the new Ferdinand
Porsche-designed 350bhp engine and this led to a
contract in October 1916 for the Rapp Motor Works
to produce 224 engines worth 10 million marks.
Because of Rapp's dubious reputation for quality, the
Austrian War Ministry insisted on appointing a
supervisor to oversee the production.

This supervisor was Franz Josef Popp, a 30-year-old
Austrian Lieutenant, and qualified engineer. Popp was
initially dismayed at the sight of only three wooden
sheds for the works, but after a week was convinced
that Rapp and his loyal workers could produce the
engines. This coincided with a dramatic turn of events.
With America's entry into the war the Prussian military
administration requested more aircraft. They didn't
want the 350bhp engines Rapp was building, but
required six-cylinder engines for fighters. Rapp's
reputation was so tainted that Popp, now technical
manager, decided the only solution to ensure survival
was to remove Rapp's name from the company. Thus,
on 20 July 1917, the Rapp Motor Works became the
Bavarian Motor Works, and on 5 October 1917, Popp
registered the rotating-propeller trademark at the
Imperial Patents Office. With the creation of BMW,
Rapp immediately left, although he had earlier retired
through illness. Despite a precarious business
environment, Popp was intent on expansion,
purchasing 25 hectares of land on Moosach Street on
the northern end of Oberwiesenfeld one month after
forming BMW. At this time, there was no merger with
the Bayerische Flugzeug-Werke (Bavarian Aircraft
Works and formerly Otto Works). This continued to
coexist happily with BMW until 1922.

On 2 January 1917, Max Friz joined Rapp Motor
Works. He had known Rapp when they worked in the
design department of Daimler together, but had
resigned from Daimler when his request for a 50-mark
pay rise was declined. Friz wrote to Rapp asking for a
job and he brought along a design for a high-altitude
carburettor that incorporated an inlet fitting that
opened at an altitude of 2,000m (6,500ft), but which
remained closed at lower levels. The dour Rapp was

sceptical, but Popp decided Friz's idea was worth investigating. When the Austro-Daimler engine contract was cancelled in May 1917, Popp sent Friz off to redesign Rapp's problematic six-cylinder engine. The result was an engine that no longer vibrated, and produced 160bhp at 3,000m (1,000ft), considerably more than the comparable Daimler unit. Friz convinced the authorities in Berlin of the advantages of his new engine, the IIIa, and by the end of 1917, had tested it successfully in the air. So superior was the IIIa that by the middle of 1918, the Prussian military ordered 2,000 engines. Popp's gamble on expansion was vindicated, but the company was soon in trouble.

It was one thing to have a superior product, but another to be able to produce it. Food and material shortages in Munich in 1918 severely affected worker morale and engine production slumped. With 1,744 workers, the costs associated with building the new premises at Moosach Street, as well as Wiedmann's extravagant lifestyle, the only solution was to convert BMW from a limited to a public company. So, on 13 August 1918, BMW became a public company, with one of the main investors the same Camillo Castiglioni. This allowed the expansion of the workforce to 3,400, and engine production to 150 per month by October 1918.

Although Ernst Udet achieved 30 victories with his BMW IIIa-powered Fokker D.VII fighter plane, and new BMW II and IV engines were about to enter production, it was too late. The end of the war arrived on 11 November and by 6 December production at BMW had ceased. Somehow, Popp managed to reopen the factory on 1 February 1919 enabling further development of the BMW IV engine, a development of the IIIa. Zeno Diemer used this engine on 9 June 1919 to reach an altitude of 9,760m (32,023ft), a world record that wasn't ratified by the international authority, the FAI, as Germany was a defeated power. Only eleven days after Diemer's repeat flight on 17 June, and just as Popp believed this record would set the company on a new path, the Treaty of Versailles was signed. Germany was now forbidden to be involved in the manufacture of aircraft and engines until mid-1920. Popp was now in a dilemma as to what BMW could manufacture.

While Friz was designing engines for boats and trucks out of the remains of aero-engines in the cellars, salvation appeared from experienced foreman, Martin Stolle. Stolle was a committed motorcyclist who had won third prize in a race from Vienna to Munich on a 1913 model English Douglas in 1914. Stolle joined BMW in 1917 and, impressed by the reliability of the Douglas flat-twin, he persuaded Popp to sanction the development of a similar motorcycle engine.

Orders for the M2 B15 were too small to sustain the company so Popp negotiated a contract with Knorr

Martin Stolle astride the Victoria KR1 with the BMW M2 B15 engine. *Mick Woollett*

THE M2 B15

Early in 1920, Stolle acquired a 1914–15 Model B 500cc Douglas motocycle and stripped it down on Friz's workbench. Every part was measured and drawn up reluctantly by Friz, an innovator used to designing every part, and loath to copy something else. From these drawings, Stolle built six test engines before installing one in the Douglas chassis and riding over to the Victoria Works in Nuremberg. Victoria then fitted the engine in their frame and marketed it as the Victoria KR1. Essentially, the M2 B15 ('2' for two cylinders and 'B' for boxer) was a copy of the 494cc Douglas, sharing the side-valve layout and 68 x 68mm bore and stroke, but with a BMW trademark on the crankcase. Friz incorporated enclosed valves and force-fed gear lubrication. With an exposed flywheel and non-detachable cylinder heads, the engine weighed only 31kg (68lb), and with a single carburettor produced a modest 6.5bhp at 3,000rpm. Positioned in the frame longitudinally like the Douglas, with either belt or chain final drive, the M2 B15 soon found its way into other motorcycles besides the Victoria. This included Corona, SMW, Bison, RS Bayerland, Heller, and later the Bayerische Flugzeug-Werke Helios, motorcycles. On 28 March 1921, Stolle entered a Victoria KR1 in the 370km (230-mile) Bavarian Motorcycle Derby.

During 1920, Bayerische Flugzeug-Werke produced the two-stroke Flink under contract for Karl Rühmer. *Stefan Knittel*

After BMW moved into the BFW premises during 1922, they inherited a number of Helios motorcycles with the BMW M2 B15 engine. *Rainer Schwinge*

THE FLINK AND HELIOS

By 1917, BFW was producing 200 aircraft a month and had a workforce of 2,400. Then, a fire in November 1917 reduced the works to ashes, from which BFW never really recovered. The buildings weren't rebuilt, and as BFW was prohibited from aircraft manufacture following the Versailles Treaty, it was engaged initially in building office furniture out of leftover wooden aircraft mountings. Shortly afterwards, an engineer, Karl Rühmer, contracted BFW to produce a simple motorised bicycle that he had designed. Rühmer undertook to manage the sales and marketing and the Flink was born. With a 1½bhp 143cc single-cylinder two-stroke engine bought from Carl Hanfland in Berlin, the Flink weighed 40kg (88lb) and had a top speed of around 50km/h (31mph). The cylinder was angled forward with fins parallel to the ground, while the magneto bolted to the back of the crankcase to

form a 'V' shape. The engine was mounted in Rühmer's simple loop frame with sprung forks and drove the rear wheel directly by belt. At the same time, Gustav Otto, BMW's founder, was also engaged in the production of a motorised bicycle. Otto's Flottweg featured a 1bhp four-stroke engine driving the front wheel through an epicyclical gear.

The success of the Flink during 1920 and into 1921 saw BFW decide to compete with the Victoria. After building a prototype Helios with a Douglas engine in October 1921, Rühmer used the BMW M2 B15 engine for production examples. The chassis of the Helios was similar to the Flink, and was not sufficient for the heavier engine. This was no match for the Victoria in the depressed days of 1922, and only 1,015 examples of the Helios were produced, the last one in November 1923.

Bremse AG in Berlin to supply 10,000 compressed-air brakes for the Bavarian Railways Board. In May 1920, Knorr Bremse bought Castiglioni's BMW shares and set up production in Moosach Street. Over the winter of 1920–21, production of the M2 B15 commenced, continuing alongside brakes. However, by 1922, the German mark was virtually worthless due to rampant inflation and the board of directors instructed Popp to concentrate purely on brake production. But Popp wasn't about to give up engine production. He already had plans to produce a small, 1,100cc car (the Sasha) and again approached Castiglioni. Castiglioni suggested they bought back the name BMW from Knorr Bremse, along with the construction facilities not required for brakes. Popp then went to Knorr with 75 million Marks and came away with the BMW name and symbol, and on 20 May 1922, the contract was

ratified. Knorr Bremse remained in Moonbach Street, the name changing to Südbremse AG in July 1922. All BMW needed now were new premises.

Coincidentally, in nearby Neulerchenfeld Street, the Bayerische Flugzeug-Werke was close to bankruptcy. Castiglioni had acquired a majority shareholding in BFW in November 1921, and on 5 June 1922, BFW became Bavarian Motor Works. Camillo Castiglioni was listed as the only shareholder, remaining on the board until October 1929 when he was forced to relinquish his shareholding under mysterious circumstances.

When BMW moved into the BFW premises they inherited the small-scale Helios motorcycle production line and a considerable number of unsold examples. Popp commissioned Friz to modify the Helios to allow the remaining stock to be sold. Although Friz was a competent and enthusiastic motorcyclist, he was reluctant to work on the Helios. However, he repositioned the engine in the frame to improve steering and handling. Popp then instructed Friz to consider designing a completely new motorcycle, the R32. The main reason for this was that there was no longer a market for the M2 B15 engine as Victoria, BMW's main customer, was now using an engine from WSM (Wilhelm Sedlbauer Munich). Martin Stolle created this engine, but had left BMW in 1922 because BMW had refused to pay 100 marks in expenses for a trip. Stolle's overhead valve engine produced 8bhp, and powered the most successful sporting motorcycle in Germany at the time, the Victoria KR2. It also proved superior to the new BMW R32 at Solitude in June 1923, where none of the three special overhead valve BMWs finished.

THE R32

Friz retired to his home at 34 Riesenfeld Street near the BMW factory, working uninterrupted from the guests' room. As BMW already had a 500cc horizontally opposed twin-cylinder engine, Friz took this as a basis for his new motorcycle. The rear cylinder tended to overheat on the Victoria and Helios, so he decided to mount the engine transversely, adding a shaft drive. The rigid frame meant the driveshaft didn't require a universal joint, and a rubber disc (Hardy) was sufficient as a shock absorber. Although the English Sopwith ABC motorcycle of 1919 also featured a transverse twin-cylinder engine (without shaft drive), according to Friz he was unaware of the ABC at the time. Later, there were accusations from Granville Bradshaw, a designer at ABC, that the BMW was a copy of their motorcycle, but there were too many detail differences for this to be substantiated. Within four months, in December 1922, Friz completed the drawings.

The side-valve engine, a development of the M2 B15, was known as the M2 B33 or M33a, and was uprated to produce 8.5bhp. While this power output wasn't particularly outstanding, it was the execution of the design which set the R32 apart. Here was a motorcycle designed with reliability and ease of maintenance foremost. The engine, including the valve timing system, was fully encased, there were no chains requiring adjustment, and there was a hand-operated side lever for a three-speed, grease-filled gearbox. The ignition was by a magneto generator and a rather complicated set of handlebar controls.

Friz installed this engine in a closed twin-loop tubular steel frame, with the petrol tank underneath the upper frame tubes. The frame was brazed and sleeved, and as the workers lacked experience in brazing, fractures on the solder joints were a problem until the introduction of pressed-steel frames in 1929.

The side-valve R32 engine was a development of the M2 B15, and every component was encased. *Cycle World*

The front suspension consisted of a short swinging fork with a cantilever plate spring beneath the steering stem. Initially, there were no front brakes, a block brake on the rear being operated by the rider's heel, but by 1925 there was also a 150mm drum front brake. The R32 ('R' for Rad meaning cycle, but the '32' remains a mystery), not only offered a top speed of around 90km/h (56mph), but the low centre of gravity promised safe and manageable handling for a touring motorcycle on the poor quality roads of the day. In May 1923, Friz himself tested the R32, finishing a club outing through the Bavarian mountains without incurring any penalties. The R32 was launched at Berlin in September 1923, one month before the Paris Car Show where it was a star attraction. It established a boxer-twin shaft-drive format that would characterise BMW motorcycles through until the present day, but the initial response was mixed. Sceptics feared the engine could be easily damaged in a fall, others felt it was underpowered, but no-one could deny the compact engine and transmission unit was a brilliant design and beautifully executed.

The release of the R32 coincided with stability of the German mark, but at 2,200 marks (or 2,600 marks with a light, horn, pillion seat and speedometer) it still represented a significant investment. However, the motorcycle market was flourishing in Germany as cars were for the wealthy few and with the demand for motorcycles strong, BMW managed to sell 1,500 R32s by the end of 1924.

One of the most important figures in the history of BMW motorcycles, Rudolf Schleicher, with his first creation, the R37. *Stefan Knittel*

Below: Schleicher's 1934 R37 has been restored and is in a private collection in Bavaria. The cylinders were different to the production version, with the valve covers held by a single bolt. *Ian Falloon*

RUDOLF SCHLEICHER

Rudolf Schleicher was born in Basle, Switzerland, in 1897 of German parentage, graduating from the Munich College of Advanced Technology after the First World War. Schleicher was an enthusiastic motorcyclist and went to work for BMW in September 1922 under Friz and Martin Duckstein. As well as working on the design of the R37 Schleicher gave BMW their first motorsport success, at the 1924 ADAC Winter Rally in Garmisch-Partenkirchen. In April 1927, following a disagreement with Friz over supercharging and the viability of two-stroke engines in cars, he went to work for Horch in Zwickau as manager of the experimental engine department. There, Schleicher was involved in the development of Horch eight-cylinder engines with Fritz Fiedler. Returning to BMW in 1931, he became chief of the experimental department for cars and motorcycles, and from 1932 was in charge of motorcycle design. Schleicher also developed BMW's first six-cylinder passenger-car engine. He remained head of motorcycle development until 1945 and at the end of the Second World War set up Schleicher Fahrzeugteile KG, repairing cars and manufacturing camshafts. The company was later taken over by his sons Rolf and Hans, with Rudolf returning to BMW AG's development and experimental department from October 1956 staying through to his retirement in 1960.

THE R37

After the humiliation at Solitude in 1923, BMW returned the following year with a revised machine designed by Schleicher. Overheating of the steel cylinder and monoblock head was the main problem with the M35a overhead valve engine and it was only reliable when the ambient temperature was extremely low. At Garmisch early in 1924, Schleicher and the M35a were successful, but it was obvious that there was no real competition future with this design.

Shortly after joining BMW, Schleicher drew up a light alloy cylinder head with fully enclosed and lubricated overhead valves set at an included angle of 90°. This was quite a radical development as the valves were usually exposed to dust and water and often required manual lubrication of the rocker shafts. To achieve improved cooling, Schleicher incorporated (and patented) circumferential cooling fins with cooling passages around the valves. The cylinders were turned from a billet of steel and as Friz was once again involved in aircraft engine design (the BMW VI V12), it was left to the young Schleicher to develop the M2 B36 (M36a) for Solitude. The chassis was similar to the R32, and the new machine was given the designation R37.

Schleicher headed a team of three at Solitude, being joined by Franz Bieber and Rudi Reich. The three new R37s produced a claimed 22hp and won three

The R37 was the first sporting
BMW, but was extremely expensive
in its day. *Ian Falloon*

categories, with Reich setting the fastest time of the day. This success continued in the Eifel Mountain Race, the German Grand Prix, the Avus Race in Berlin, while Bieber took the German Road Championship. The R37 then went into limited production during 1925, albeit at a price of 2,900 marks, making it the most expensive German motorcycle available. With a new, three-slide 26mm BMW carburettor the power was 16hp at 4,000rpm, enough to propel the R37 to 115km/h (71mph).

It wasn't until 1926 though that the R37 managed to capture the attention of those outside Germany. Schleicher and his friend Friz Roth ventured to England on production R37s as private entrants in the Six-Day Race. Not realising that off-road tyres were necessary, and unable to source any, they fronted with conventional tyres and low pressures. Almost laughed out of the race when they arrived, Schleicher crossed the finishing line with a gold medal. The British press

sat up and took notice, and Professor A. M. Low wrote in the Auto-Cycle Union magazine: 'After the toughest days in the field we could not find a single oil leak, the machine was beautifully quiet, and seemed to possess great reserves of power. From a design point of view it is miles ahead of any British machine.'

THE R42 & R47

The R32 established BMW as a manufacturer of quality motorcycles, but after two years a replacement was needed. At the end of 1925, a redesigned touring twin appeared alongside the R32, and replaced it during 1926. Although still a side-valve 500, the R42 incorporated a number of developments, including a wedge-shaped combustion chamber to improve combustion. Designated the M43a, there were new cylinders with the cooling fins set across the barrel, and the beautifully sculptured alloy cylinder heads were detachable. With a BMW Special two-valve 22mm carburettor the power was increased significantly, to 12bhp at 3,400rpm. The three-speed gearbox also incorporated a speedometer drive, but the top speed of 95km/h (59mph) represented no improvement over the R32.

Accompanying the engine development was a new frame with straight downtubes placing the engine further rearwards for better weight distribution. This improved the rather top-heavy feeling of the R32 and led to more secure handling. The heel-operated driveshaft brake that appeared on the R39 single also featured on the R42. There was now a sidecar mount offered on the rear driveshaft housing and the overall weight of 126kg (278lb) was up slightly on the R32. By 1926, BMW was expanding, and a reduction in price of the R42 to 1,510 marks saw it as one of BMW's most successful motorcycles of the 1920s.

A year later, an overhead valve version, the R47, replaced the R37. Not purely a racing machine like its predecessor, the R47 used an identical chassis to the touring R42, complete with sidecar option. Developments of the engine, designated the M51a, included cast-iron, rather than steel, barrels, and valve covers retained by a single bolt. The three-slide carburettor made way for a standard BMW type and the power increased slightly to 18hp. The top speed was a little less than the R37, at 110km/h (68mph), but because the price was reduced significantly, to 1,850 marks, 1,720 R47s were produced in one series. A sports variant of the R47, with an Sp.M51a engine, was also available, providing 31bhp.

Below: The R42 replaced the R32 during 1926. The side-valve engine featured a new cylinder head design and longitudinal fins. *Brian J. Nelson*

Opposite: In addition to the R42, there was a sporting overhead valve R47 for 1927. This replaced the R37. *BMW Mobile Tradition*

THE R52, R57, R62, & R63

By 1928, BMW was established as a premier manufacturer of touring motorcycles and four new models were released using the same chassis (designated the F56a). The two 500cc models were joined by 750cc versions, but these were new designs with a pressed-up crankshaft instead of the earlier one-piece type. There was also a redesigned lubrication system and the side-valve engines now featured a longer, 78mm stroke, while the overhead valve engines retained the earlier, 68mm stroke.

The R52 touring side-valve 500 (designation M57a) had a 63 x 78mm bore and stroke, with the side-valve 750 R62 (M56a) a square 78 x 78mm. There was a new mounting system for the side-valve cylinder heads but they were visually similar to before. The overhead-valve R57 (M59a) 500 remained square at 68 x 68mm, with the overhead-valve R63 750 (M60a) featuring almost modern oversquare dimensions of 83 x 68mm. The power of the R52 and R57 was the same as the earlier R42 and R47, while the 750 R62 produced 18bhp and the R63 now gave 24bhp at 4,000rpm. This appeared to be very conservative as the R63, with its equally modest claimed top speed of 120km/h (75mph), was one of the fastest machines available at the time. It also came at the premium price of 2,100 marks so it wasn't surprising that only 794 were produced. As with the R47, there was also a sporting version of the R63 available. With an Sp.M60a engine, the power was 36bhp. BMW also produced a 40bhp 749cc racing machine during this period.

All four new models came with a magneto-generator electrical system, with optional Bosch lights until 1929 when they became standard. There was a stronger three-speed gearbox with oil instead of grease, and a new hand-change mechanism. Underneath the gearbox was a useful toolbox, with a hinged door. Also for the

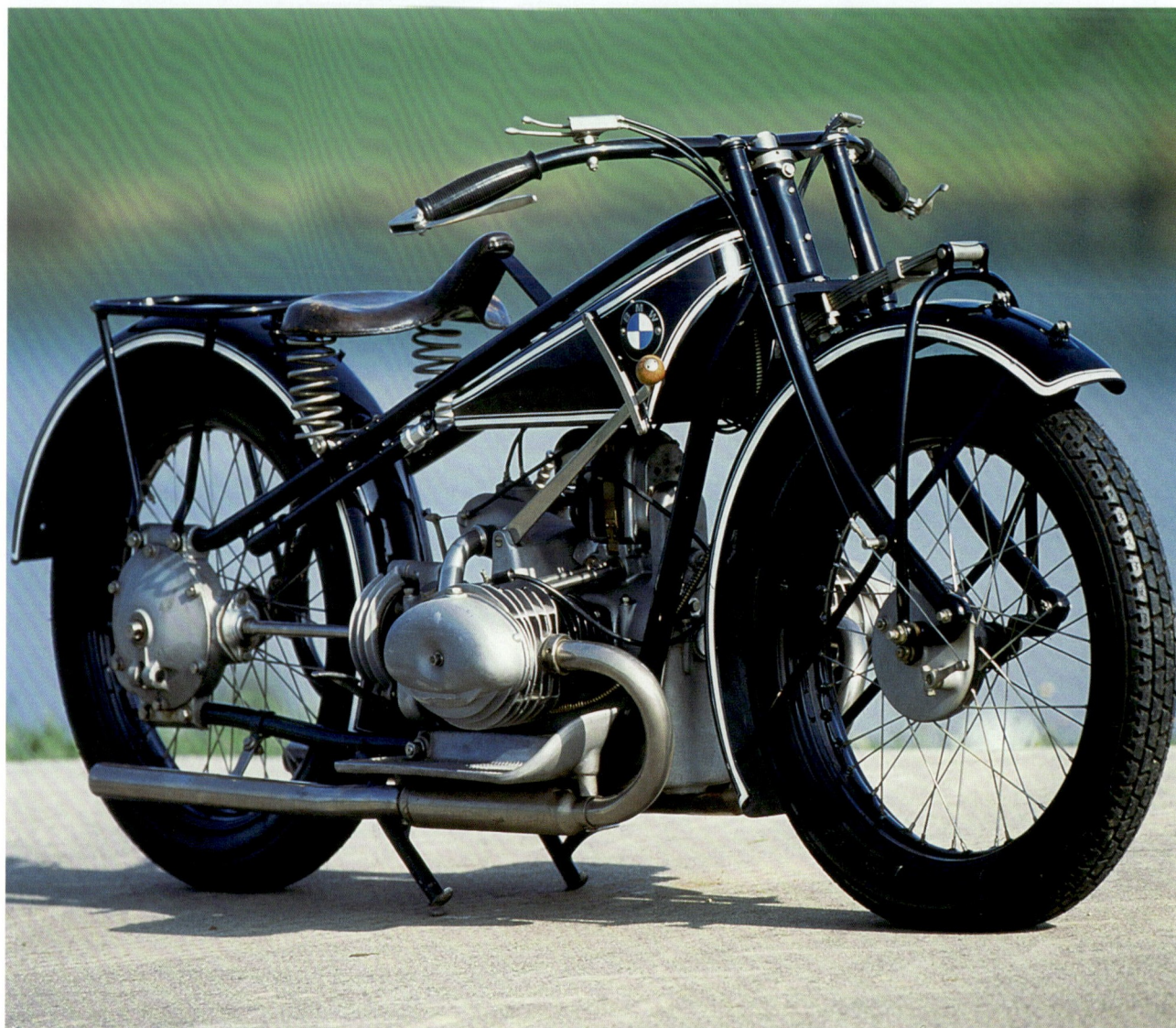

first time, the kickstart was side-mounted, kicking out to the side, which was to be a BMW feature for many years to come. The clutch was also originally a single-plate but was changed during production to a twin-plate. The F56 chassis followed the form of the R42 and R47, but with a larger front brake (200mm) and a six-leaf spring for the front forks, rather than a five-leaf. The weight was increased considerably through, and even the R52 weighed 152kg (335lb). All these models were distinguished by the triangular fuel tank mounted between the engine and the top frame tubes, with the majority being side-valve examples.

As the BMW still lacked the handling finesse of the British competition, racing versions were supercharged from 1928. Supercharging was initially instigated by Rudolf Schleicher during 1927, and was continued by Friz after Schleicher left BMW. Friz installed a French Cozette centrifugal supercharger, and later a Zoller rotary-vane type, mounted horizontally above the engine and gearbox where it was driven by an oil-bathed chain from the crankshaft. This was extremely effective, providing the racing R57 with around 55bhp and the R63 with 75bhp, and during 1929, the racing BMWs were virtually unbeatable in German competition. Hans Soenius won three German championships and Ernst Henne was another of BMW's outstanding riders. Convinced he could also take the world speed record away from the British riders Oliver Baldwin and Bert Le Vack, Henne persuaded Friz to prepare a short-stroke supercharged 750. On 19 September 1929, on the narrow tree-lined Ingolstädter road near Munich, Henne raised the absolute motorcycle world speed record for the mile to 134.78mph (216.9km/h). The BMW was unfaired but Henne wore a streamlined helmet and tail attached to his riding suit. Because the FIM refused to acknowledge his kilometre record, as he should have attempted this before the mile, Henne returned a few days later and did it again.

Henne's records set off a competition between BMW, Brough Superior, and Gilera for the world speed record that would last throughout the next decade, and were the culmination of an extremely profitable period for BMW. Not only were nearly 20,000 motorcycles produced during the 1920s, but also aircraft engine manufacture was well underway again. Production of the type VI V12 began during 1926 and 7,000 units were sold by the mid-1930s. In 1928, BMW signed a licence agreement with the American company Pratt & Whitney to produce radial aircraft engines.

Perhaps the most significant development during 1928 was the purchase of the Eisenach car plant near Frankfurt, along with the licence to build a copy of the British Austin Seven called the 'Dixi'. BMW then also became a car manufacturer, with vehicles manufactured at Eisenach throughout the 1930s while motorcycle and aircraft engine production remained at Munich. Despite this new emphasis, motorcycle production remained pivotal to the company, and during 1929, BMW released two new 750cc models with totally revised frames. These were the R11 and R16, machines that would serve until 1934.

Opposite top: The BMW twin grew to 750cc with the R62 of 1928. *BMW Mobile Tradition*

Opposite bottom: One of the fastest motorcycles available at the end of the 1920s was the short-stroke overhead valve 750cc R63. *BMW Mobile Tradition*

Right: Ernst Henne gave BMW their first world speed record in 1929 on a supercharged 750, but also set 500cc records. This is in April 1932, on the way to setting a new flying kilometre record. *Mick Woollett*

After the war, Meier resurrected the
500 Kompressor, and raced it
in German events until 1951.
Ian Falloon

2 HIGHS AND LOWS:
THE 1930s, THE SECOND WORLD WAR, AND POST-WAR RECONSTRUCTION

Ernst Henne ushered in the new decade with his world speed record, but this was short lived and he lost it soon afterwards to Joe Wright on the JAP-powered OEC Temple. Henne then set about regaining it, and on 30 September 1930, raised the record to 221.539km/h. Records were significant, both for BMW's importance as a world marque, and for national pride. Economically this was a difficult time, and motorcycle sales were directly related to Henne's achievements. After losing the record again to Wright, who raised it to an astonishing 242.568km/h (390.292mph) in November 1931, at Cork in Ireland, Henne attempted to regain it during 1932. After several failed attempts on the Neunkirchener Allee, a long straight road south of Vienna, there was a slump in motorcycle sales. Production slipped from 6,681 in 1931 to 4,652 in 1932.

Salvation for Henne and BMW came with the return of Rudolf Schleicher. During 1930, Henne persuaded Popp to entice Schleicher back from the troubled Horch company, and with Sepp Hopf, Schleicher designed a new multi-plate supercharger. On 3 November 1932 in Tata, Hungary, in front of a full military line-up and the Governor of Hungary, Admiral Horthy, Henne beat Wright's record and achieved 244.399km/h (393.238mph). Two years later, in October 1934 in Gyon, Hungary, Henne went slightly faster at 246.069km/h (395.925mph), and in 1935 gave the supercharged ohv 750 its final record. On the new Frankfurt–Darmstadt autobahn he reached 256.040km/h (411.968mph), and things were looking up, again.

THE R11 & R16

Although the four-model range released in 1928 was selling well, problems with frame fractures and collapsing front forks (particularly when adapted for a sidecar) prompted the release of two new 750s for 1929, the R11 and R16. The side-valve engine of the R11 was carried over from the R62, while the R16 used the overhead-valve R63 engine, and apart from having larger carburettors, these were unchanged. It was the riveted pressed-steel frame that set the new machines apart, and although undeniably stronger, the frame was also heavier, and in the eyes of many, was ugly, scarring Friz's creation.

Rudolf Schleicher and Ernst
Henne, the two main
proponents in the setting of the
series of motorcycle world speed
records during the 1930s. *BMW
Mobile Tradition*

ERNST HENNE

One of BMW's most successful riders, Ernst Henne, was born in 1904 in Weiler (Allgäu). At the age of 15 he became an apprentice motor mechanic in Ravensburg, and entered his first race riding a Megola in Mühldorf, Bavaria, in 1923. In 1925, he switched to an Astra, and after taking sixth place in the European Grand Prix, at Monza in 1926, was signed up as a factory rider by BMW. He was 500cc and 750cc German champion, in 1926 and '27 respectively and also won the 1928 Targa Florio in Sicily, all on supercharged machines. From 1932 to 1936 Henne was a member of the German national team for the International Six-Day Trial in England; the first win coming in 1934. Henne also raced cars during the 1930s, and was badly injured in a Mercedes in testing on the Nürburgring. Later he drove the 2-litre BMW 328 sports car to victory in the 1936 Eifel Race and again in the 1937 Grand Prix des Frontières in Chimay, Belgium. Henne then pursued a successful business career as a Mercedes dealer in Munich, but is best remembered for his 76 motorcycle speed records for BMW. He was a tough man and in 1929 he narrowly escaped death when the front wheel spindle came out of the forks while he was travelling at over 130mph (210km/h). Managing to bring the machine to a standstill Henne later commented: 'Fate was kind to me that day.' A few months later, early in 1930, he went to Ostersund in Sweden with spiked tyres and managed 198.2km/h (123.2mph) for the kilometre on a frozen lake in minus 14°C temperatures. Henne set this speed immediately before crashing and sliding 500m (550yd) along the ice.

The F66 chassis of the R11 and R16 consisted of two loops in a single pressing, joined by cross members. These strengthening sheets were also riveted at the front and to the fork blades. The only welding was at the front where the two halves joined together over the steering head. The trailing link front forks were also pressed steel, with nine-leaf-spring front suspension. The fuel tank was almost hidden by the heavy gusseting around the steering head. The result was a machine which conveyed a solidity and robustness that appealed to commercial and military interests, but was hardly a sporting mount. The weight was around 10kg (22lb) more than the R62 and R63, and while the R16 was capable of 120km/h (75mph), the R11 struggled to better 100km/h (62mph).

After production was delayed during 1929 due to some front fork problems, the pressed-steel models replaced the tubular steel framed versions in the following year. Only the sporting overhead-valve 500cc R57 remained through 1930, while the side-valve R52 disappeared altogether. The R11 and R16 were then developed through five series until 1934.

SERIES 2

The Series 2 R11 and R16 appeared during 1930. The engine was designated the M56 S2 for the R11 and M60 S2 for the R16. A stronger thrust bearing was added to the twin-plate clutch, there was an additional bearing in the rear drive, and the rear driveshaft brake shoes were increased to 55mm (from 37mm). For this series, the Bosch headlamp changed from the older drum style to a more modern cup shape.

SERIES 3

New carburettors appeared on the M56 S3 and M60 S3 engines for 1932. The R11 carburettor was now a SUM CK3/500 F1 from Berlin, with pre-heated secondary air drawn through a tube on the exhaust manifold. On the R16 the compression ratio was increased to 7:1, and twin one-inch Amal type 6/011 carburettors (made under licence by Fischer in Frankfurt) bolted directly on the intake manifolds of each cylinder head. This was enough to see the power increase quite dramatically to 33bhp and the top speed to 126km/h (78mph).

Above: The pressed-steel frame of the 1936 sporting R17 may have been old fashioned, but the telescopic front forks were innovative. The heavy styling mimicked the contemporary German cars. *Ian Falloon*

Left: Although it still used the long-stroke 750cc side-valve engine of the R42, the R11 featured a new pressed-steel frame and trailing link forks. *BMW Mobile Tradition*

SERIES 4

With the M56 S4 and M60 S4 engines for 1933 came new, single-row caged roller big-end bearings instead of the earlier twin-row rollers. The gearchange mechanism was moved to a gate underneath the knee rubber on the right, while the saddle now had extension springs. Battery ignition was also offered on the R16.

SERIES 5

For the final series of 1934, the M56 S5 and M60 S4 received a roller timing chain instead of gears, to drive the camshaft from the front of the crankshaft. The M56 S5 featured battery and coil ignition for the first time on a BMW motorcycle. There was also a revised and more effective fishtail silencer and the R11 now had twin Amal carburettors (6/406 SP and 6/407 SP). The power rose to 20hp, but there was also an R11/5 RW single-carburettor version specifically for the army. A few R11/6 motorcycles were also built with a three-shaft four-speed gearbox as a precursor to the R12.

THE 500 KOMPRESSOR

In January 1933, Hitler became Chancellor of Germany and the Nazis were well aware of the boost to morale associated with competition success. As part of their nationalistic propaganda programme they encouraged German manufacturers to embark on racing, which resulted in a new, 500cc supercharged BMW twin that would eventually inspire a range of production models.

Instead of developing the existing overhead valve design, the 500 Kompressor was a purpose-built grand prix racer. Back in 1928, drawings were made for an overhead camshaft engine (M61a), and considerable time was spent investigating bevel-gear-driven overhead camshaft systems, through until 1932. There were four versions: 500cc (M250/1, M255/1) and 600cc (M260/1, M265/1) with or without a supercharger. The bore and stroke of the 500 was 66 x 72mm, and 72.2 x 72mm for the 600, with the valves operated by bevel-gear-driven twin-overhead camshafts in the cylinder head. Each pair of camshafts was geared directly to each other and opened the valves through short rockers. The Zoller multi-cell vane-type supercharger was now spline-driven from the front of the crankshaft, with a single 27mm Fischer-Amal side-mounted carburettor on the right.

Where the new Kompressor was superior to earlier examples was in the design which now included long induction ports under each cylinder to the rear intakes that ensured the mixture was adequately cooled. Because the supercharger ran at engine speed it

Wiggerl Kraus debuted the 500 Kompressor at Avus in 1935, and during 1936 rode it to victory in German events. *Mick Woollett*

provided boost of around 15psi, and its position improved weight distribution. The power delivery was very smooth, and the Kompressor would pull cleanly from as low as 2,500rpm. Many of the castings were lightweight electron magnesium, including the crankcases and gearbox housing and for the first time for BMW, there was a four-speed gearbox with a positive foot gearchange.

Rather than the pressed-steel frame of the production models at the time, the Kompressor used Schleicher's electrically arc-welded tubular steel frame, and incorporated his own design of oil-filled 28mm telescopic forks. This was the first time oil-damped telescopic forks appeared on a motorcycle, but the rigid rear end remained.

The Kompressor debuted at the high-speed banked Avus circuit near Berlin in June 1935. Ridden by Ludwig 'Wiggerl' Kraus, it was beaten by Ragnar Sunqvist on the Swedish Husqvarna V-twin at an

The rigid-framed Kompressor was
also used for the ISDT of 1935 and
1936. *Mick Woollett*

Henne on the Frankfurt autobahn
in November 1937, about to set
another world speed record.
Mick Woollett

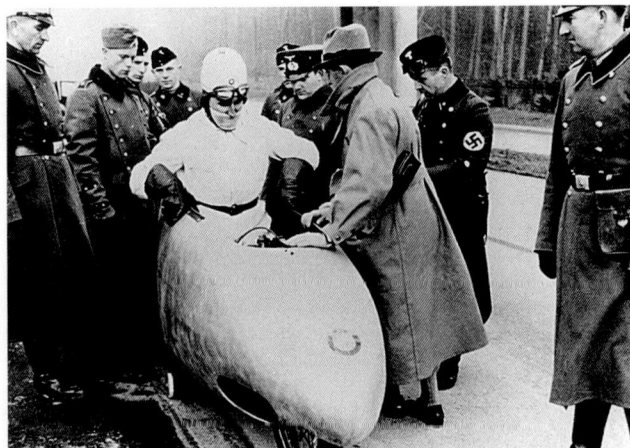

average speed of 170km/h (106mph). There were no
more outings that year except for the International Six-
Day Trial where the German trophy team rode detuned
Kompressors to victory.

For the 1936 season, Otto Ley and Karl Gall
received works Kompressors, Ley taking second in the
Swiss Grand Prix behind Jimmy Guthrie on a Norton.
Despite its superior power, the Kompressor was a
handful and difficult to ride. Ley managed another
second at Assen before the Kompressor achieved its
first victory in Sweden, on 30 August. Ley and Gall
finished first and second ahead of the Norton, FN, and
DKW works teams.

In the meantime, the world speed record was lost to
Eric Fernihough on the Brough Superior, so Schleicher
prepared a fully streamlined 500cc Kompressor. In
October 1936, on the Frankfurt–Darmstadt autobahn,
Henne went at 272.006km/h (169.025mph). BMW
now looked optimistically towards the 1937 season,
along with Alexander von Falkenhausen's new rear
suspension. Faced with reluctance by the riders to race
with, Falkenhausen had proved the superiority of his
rear suspension by successfully riding the machine
himself in the International Six-Day race at Füssen. The
plunger rear suspension featured straight guide sleeves
in vertical tubes, and the driveshaft required a universal
joint.

In the hands of Ley and Gall, the fully sprung
Kompressor now proved a match for the British
machines. Gall outpaced the Nortons at the Dutch TT
to take the victory, also winning the German Grand
Prix at the Sachsenring, after Guthrie was killed on the
last lap. Ley won the Swedish TT and Jock West the

Ulster, giving BMW four out of seven Grand Prix
victories in 1937. Jock West also rode a solitary
Kompressor at the Isle of Man, finishing a creditable
sixth, and Gall won the German Championship. The
year ended with Henne responding to Fernihough and
Piero Taruffi's new world speed records. With wind
tunnel-tested streamlining, on 28 November 1937,
Henne achieved 279.503km/h (173.683mph) on the
90bhp 500cc. The record would last until 1951 as
Fernihough was killed at Gyon in April 1938 trying to
beat it.

The Isle of Man was the first event for 1938 and
BMW fielded a three-man team. Developments saw the
power up to 55hp at 7,000rpm running on petrol-
benzol, and larger, full-width brakes. With the Nazi
decree that a German should win on a German
machine, Georg (Schorsch) Meier replaced Ley,
alongside Gall, while West filled out the team. It was
West who was fastest, while Gall crashed during
practice and Meier was unable to remove the soft
warm-up spark plugs on the start line. West finished
fifth, Harold Daniell winning on the works Norton.
Meier than went on to win the Belgian Grand Prix, the
Dutch TT, and the German Grand Prix to take the
European Championship. West again won the Ulster
Grand Prix.

With war clouds looming, Norton withdrew from
racing for 1939, leaving BMW to battle with Gilera
for 500cc honours. Meier, Gall, and West went to the
Isle of Man in June, Gall crashing at Ballaugh Bridge
during practice, dying four days later. Despite this
setback, Meier won the Senior TT at 89.38mph
(143.81km/h), with West second, more than two

Schorsch Meier on his way to
winning the 1939 Senior TT at
the Isle of Man. *Mick Woollett*

minutes behind. Meier followed his TT victory with
wins at the Dutch TT and the Belgian Grand Prix.
Here, he lapped at 161.96km/h (100.63mph), the first
time a 'ton' lap was achieved in a classic event. A crash
in Sweden saw Meier out of the German Grand Prix
due to injury and the outbreak of war ended the season
after Ulster. The final Kompressor was a formidable
machine, if somewhat of a beast to ride. When weighed
after the TT, Meier's machine was found to be the
lightest finisher at only 137kg (302lb).

During the Second World War, all the factory racing
machines were transferred to Berg on Lake Starnberg
for safe keeping, but Meier managed to retrieve the Isle
of Man machine in 1943, having hidden it away in a
hay barn on the Karsfeld estate at Allach. When he
wheeled it out for a demonstration race against the

NSU at Solitude in 1947 hundreds of thousands of
spectators turned out to watch. After this, the 500
Kompressor (M250/2) was further developed. There
were redesigned superchargers (some two-stage),
aluminium cylinders, and lighter reciprocating parts,
producing as much as 95bhp on occasion.

THE R12 & R17

Motorcycle production more than doubled in 1934,
from 4,734 in 1933 to 9,689, encouraging the
development of two new 750cc models. These were the
side-valve R12 and overhead-valve R17, which were
first displayed at the Berlin Motor Show in February
1935. The engines were based on the earlier R11 and
R16, but with a four-speed gearbox, and while the

The side-valve R12 with a single
carburettor was extremely
popular with the German military,
and was produced until 1942.
Brian J. Nelson

pressed-steel frame was retained, what set the new
machines apart from earlier models, were Rudolf
Schleicher's oil-damped telescopic forks. These had first
appeared on Alfred Böning's R7 prototype of 1934 and
were the first modern-style hydraulic forks fitted to a
motorcycle. The R12 and R17 were a curious
combination of the old and the new, retaining a rigid
rear end when many British motorcycles featured rear
suspension. Even Hitler expressed his surprise at this as
he passed the BMW stand at the 1935 Berlin Motor
Show when he asked Schleicher: 'and when are we
going to get rear suspension?' Schleicher later admitted
he was filled with embarrassment and consternation by
Hitler's question, but it would still be two years before
rear suspension appeared. Schleicher already had
Böning's rear suspension design based on the Norton,

but he was unhappy with it. He then had Falkenhausen
design a new system, with sliding tubes housing the
driveshaft with springs at the frame ends.

The R12 and R17 four-speed engines featured
stronger crankshafts, but retained the hand-change
through a gate on the right side. The R12 (M56 S6 or
212) also came with a choice of a single Sum CK
25mm carburettor or two Amal 6/406/407 carburettors
and the power was identical to the two versions of
R11. As the flagship of the range, the R17 (M60) only
came with twin Amal 76/424 carburettors, but revved
up to 5,000rpm to produce 33bhp. This was enough to
provide an impressive 140km/h (87mph), but the F66
chassis was still the same as the R12 workhorse,
suitable for military and sidecar duties. The rear brake
was now a 200mm drum instead of the driveshaft type,

Schorsch Meier and the
500 Kompressor reunited at
the Salzburgring in 1974.
Mick Woollett

GEORG 'SCHORSCH' MEIER

After joining the Bavarian State Police in 1929, Georg, known as 'Schorsch', Meier soon attracted attention for his ability on a motorcycle, and the fact that he rode too fast on the Bavarian country roads. However, these roads were similar to those of the Isle of Man so this stood him in good stead ten years later. By 1933, Meier was a member of the police team for the 2,000km (1,242-mile) cross-Germany run, immediately setting the fastest time in his class. Along with Josef Forster and Fritz Linhardt, he made up the successful 'cast-iron' cross-country team riding BMW R4s.

Born in 1910, Meier grew up in Mühldorf am Inn, where he was an apprentice motor mechanic from the age of 12. He always wanted to be either a racing motorcyclist or a policeman and when the Bavarian State Police force was wound up in 1935, Meier joined the German Army. This allowed him to continue riding in competitive events. He was German Army Champion in 1935 and 1936, and substituted for an ill Henne in the 1937 International Six-Day Trial at Donington in England. Meier won a gold medal, earning a ride in the BMW team. After his successful racing year on the 500 Kompressor during 1938, Meier was approached by Auto Union to race cars. Following the death of their leading driver, Bernd Rosemeyer, Auto Union needed a German driver so Meier came to an agreement with Schleicher to race cars and motorcycles on alternate weekends. At the Belgian Grand Prix he vainly tried to save Popp's son-in-law Dick Seaman from his burning Mercedes, while Meier's best result was a second in the French Grand Prix. Because he spent seven months in hospital after the Swedish motorcycle Grand Prix, Meier ended as a

driver to Admiral Canaris and head of the Abwehr (Military Intelligence) transport section during the war. From June 1945 to March 1948, he was a plant security officer in Allach, the US Army's Karsfeld Ordnance Depot. Meier then set up a motorcycle dealership in Munich, continuing to successfully race the Kompressor in German events and winning five German Championships between 1947 and 1953.

enabling the front and rear 19in wheels to be interchangeable. Although they were revolutionary for 1935, the telescopic forks were decidedly underdeveloped with only meagre one-way damping and 75mm of movement.

It was also difficult to disguise the heavy, pressed-steel frame, and for 1936, the R12 and R17 received sweeping mudguards imitating the styling of contemporary luxury German cars. While the R12, at 1,630 marks, would become the most popular pre-war BMW motorcycle, the princely sum of 2,040 marks for the R17 meant it was for the fortunate few. The most expensive German motorcycle available in its day, only 434 were produced. If viewed as a comfortable touring machine for straight smooth roads, rather than as a

sporting motorcycle, the R17 was successful. It epitomised the best German attributes: solidity, and efficiency, but by 1936, the time was right for a completely new machine. This was the R5, which not only looked much more modern, but was significantly lighter than the R17, and cost only 1,550 marks. BMW had made their sporting flagship obsolete.

THE R5 & R6

When the R5 was released at the Berlin Motor Show in February 1936, it drew heavily on the 500 Kompressor racer, and was the most advanced motorcycle available at that time. Designed by Leonhard Ischinger, it was

One of the finest motorcycles produced during the 1930s was the overhead valve R5. This 1937 version has the air filter incorporated in the gearbox housing. *Ian Falloon*

the first 500cc ohv production BMW since the demise of the R57; the engine (designation M254/1) was all-new, with the crankcase a one-piece tunnel-type similar to that of the singles. The crankshaft was inserted from the front and there were now two chain-driven camshafts instead of one, positioned over the crank to allow for shorter tappets and pushrods. The timing chain also drove the Bosch generator on top of the crankcase, with the ignition coil and distributor positioned inside the front cover. The included valve angle was reduced to 80°, and there were now double hairpin valve springs to provide safety at higher rpm. The rocker arms pivoted in needle roller bearings.

The four-speed gearbox was foot-operated by a linkage on the left, although the right-hand lever was retained primarily as a quicker way to select neutral. Carburation was by twin Amal 5/423 carburettors, each with a small ear-type air filter. These proved unsatisfactory and for 1937, were replaced by a central wire mesh air filter built into an extension of the gearbox casing. Although the power output of 24bhp at 5,800rpm was less than that of the R17, the R5 was a much more sporting motorcycle because of its electrically arc-welded tubular-steel duplex frame, similar in design to that of the 500 Kompressor. Schleicher used the same oval-section conical tubing and not only did the frame impart a more modern appearance, but the weight of the R5 was a svelte 165kg (364kg).

Completing the improved chassis specification was

Apart from the sprung rear end, the R61 was very similar to the R6. *Brian J. Nelson*

an external damping adjustment for the telescopic forks. It was only the rigid rear end that limited the ride quality, but this was compensated through a softly sprung Pagusa rubber seat. The R5 provided exceptional sporting performance for the mid-1930s with a top speed of around 135km/h (84mph) and many enthusiasts rated the handling of the rigid-frame R5 superior to the later R51. The R5 was a milestone motorcycle for BMW, finally challenging the British in performance and handling. It also provided the basis for BMW twins for the next 20 years.

Joining the R5 for 1937 was a side-valve 600, the R6. Side-valve engines were considered better suited for motorcycles fitted with sidecars and BMW were optimistic for sales of the new R5 and R6 to the German military. There optimism proved unfounded, as the German army authorities were more interested in the proven, pressed-steel frame R12. The chassis of the R6 was identical to that of the sporting R5, but the engine (designation M261/1) was new. Instead of twin camshafts, there was single, central camshaft driven by spur gears as in the earlier engines. With twin Amal M75/426/S carburettors the power was an unremarkable 18hp and the top speed was barely 125km/h (78mph). Even the good torque of the long-stroke motor that made it suitable for sidecar use couldn't save the R6 which only lasted in production for one year.

THE R51, R61, R71 & R66

As the works Kompressor racers successfully used Falkenhausen's plunger rear suspension during 1937, the release of a new range of fully sprung models at the Berlin Show in February 1938 wasn't unexpected. The R51 replaced the R5, and the R61 the R6, while there were two new models: the R66 and R71. The R71 was a 750cc side-valve sidecar machine, ostensibly to replace the R12 that was by now only a single-carburettor version and produced purely for the military. The 600cc ohv R66 assumed the position as the top-of-the-range sportster, with a list price of 1,695 marks.

Apart from a slightly lower compression ratio for the R61 there were few changes to the engines for the R51 (254/1) and R61 (261/1). As chrome was in short supply, and earmarked for gun barrels the mufflers, were generally painted black. The plunger telescopic rear suspension set the new machines apart, all sharing the 251/1 chassis, accompanied by an increase in

Above: The final side-valve BMW motorcycle was the 750cc R71. This 1938 model has black-painted mufflers, indicating a shortage of chrome at that time. *Brian J. Nelson*

Right: The sporting R66 of 1938 combined the narrower side-valve 600cc single camshaft crankcases, with overhead valve cylinder heads. *Brian J. Nelson*

weight to 182kg (401lb) for the R51 and 184kg (406lb) for the R61.

The R71 engine (designation 271/1) was based on the R61, the capacity increase to 745cc being achieved through a bore increase to 78mm. BMW's last side-valve model, with twin Graetzin G 24mm slide-type carburettors the R71, produced 22hp at 4,600rpm. Setting the R71 apart from the smaller R61 were the older style cylinders heads, but otherwise it looked identical. Undoubtedly the most exciting of the new models was the R66. Instead of using the R5/R51 twin camshaft engine with its long camchain as a basis, the 600cc R66 engine (266/1) featured the side-valve crankcases with one central gear-driven camshaft and a wider cylinder base to incorporate the pushrod tubes. The cylinder heads with hairpin valve springs came from the R51, but unique to the R66, were cylinders and heads tilted 5° forwards to provide more foot room. With a 69.8mm bore and larger, Amal 6/420 S

carburettors, the power was exceptional for an ohv twin at the time; 30bhp at 5,300rpm was enough to propel the R66 to 145km/h (90mph), and even with a sidecar it was good for 115km/h (71mph). Production of the four models lasted well into the war years, finally ending in 1941 although the last R51 appeared in 1940.

During 1939, new, 350cc (M231/1) and 500cc (M252/1) twins were considered, but development was halted in preference to the R75. By 1942, these projects were reinstated, and now incorporated many features from the R75, such as the split-valve covers. The 350 had a bore and stroke of 60 x 61mm and both versions featured battery and coil ignition. The generator, single carburettor, and airbox were all enclosed in a cover above the engine, but these models never made it to the production line.

THE MILITARY R75

As the German Army rolled relentlessly across Europe in 1940 their principal BMW motorcycle was still the rigid pressed-steel frame side-valve R12. Production of the R12 continued through until 1942, but as early as the winter of 1937–38 both Zündapp in Nuremberg and BMW were commissioned to design new 750cc military motorcycles. BMW decided to adapt the side-valve R71 engine with a split bolted tubular steel frame to allow easy engine installation and removal, with a rigid rear end. Zündapp developed their KS750 and this proved superior to the BMW. Designated the R72, the 800cc side-valve engine overheated at slow speeds, and a fan-cooled version was tested. Although BMW then considered licensed production of the

Zündapp, during 1939 they undertook development of a new design, the R75.

The most important design features of the R75 were its suitability for sidecar use, and the ability to sustain a marching speed of 3km/h (2mph) without overheating. Thus, it included sidecar wheel drive, a locking differential, and cross-country and reverse gears. The engine (designation 275/2) had overhead valves with the camshaft, a Noris generator, and a magneto driven by aluminium gears. Underneath the two-piece rocker covers were separately bolted rocker posts, unlike earlier twins with integrally cast posts. Twin Graetzin Sa 24mm carburettors fed the engine, with the air cleaned by a single moist felt air filter, with an oil strainer and sump pre-filter to ensure no dust entered the engine. This was initially positioned above the gearbox, but during June 1942 (after number 757201) it was moved to the top of the fuel

As head of motorcycle development, Alexander von Falkenhausen was involved in testing the R75. Here, he is with Hans Sachs at Groflen Donbogen during 1941.
BMW Mobile Tradition

Only a few R51RS catalogue racers were produced in 1939, and they were the closest a privateer could get to the factory Kompressor. *BMW Mobile Tradition*

THE R5SS, R51SS AND R51RS

While the double overhead camshaft 500 Kompressor always remained a factory racer, in 1937 a small number of R5SS (Super Sport) production racers was made available for selected riders. Although not offered to the public, the R5SS was essentially a modified R5, without lights or mufflers, and outside rather than inverted handlebar levers. The power output was around 4bhp more than the R5, achieved through different valve timing, polished ports and stronger valve springs, and carburettors with velocity stacks. The top speed was around 160km/h (99mph). Prior to the release of the sprung-frame R51 there was also an R51SS introduced during 1937, limited production of which continued into 1938, and a higher performance R51RS for 1939. The R51SS featured a special gearbox with higher ratios, a higher compression ratio of 8:1, and 6/432 Amal/Fischer 24mm carburettors. The power was 28bhp and the R51SS retained head and taillights. With a similar tank and seat to the works racers, the R51RS (Rennsport or Racing Sport) had 21in and 20in wheels and brakes with stiffening ribs. Although the engine was based on the pushrod R51, the long camshaft timing chain was replaced by spur gears and the cylinder barrels were the R66 type. With an output of 36bhp, the R51RS was capable of around 180km/h (112mph). Only 17 of these machines were produced.

tank, underneath a metal helmet-like cover. The power was 26bhp at 4,000rpm, with the emphasis on low-end torque, and the magneto provided automatic ignition advance. To overcome cooling problems in North Africa, Schleicher also developed a fan-cooled engine during 1942.

The four-speed transmission included a dog clutch and four lower ratios for off-road use, and the power-dividing crown wheel differential at the rear equalised any varying speeds of the two driven wheels. This allowed the R75 to perform as well as a four-wheeler and reduced tyre wear. The rear wheels, on stub axles, used the same, 4.50 x 16in tyres as the VW Kübelwagen and featured hydraulic brakes. Up front, the double-action hydraulic telescopic forks were a strengthened version of earlier BMW types, and the frame (275/1) was a lattice-girder type with a strong central box-section that could be dismantled into individual parts for easy repair.

Another design criteria for the R75 (and Zündapp KS750) was load capacity. Experience with the R12 in field conditions led to the Wehrmacht requesting a load

By 1942, the R75 was in active service, as here in North Africa with the Afrikakorps. *BMW Mobile Tradition*

of 500kg (1,100lb), corresponding to the weight of three soldiers and their equipment. However, the tyre suppliers set the maximum load at 270kg (595lb), well under the Wehrmacht's requirements. To avoid any bureaucratic problems there were two maximum total weights stated for the R75; an official figure of 670kg, and a Wehrmacht figure of 840kg (1,852lb). Most R75s were overloaded and rear tyre life was only around 3–6,000km (2–4,000 miles). As tyre supply was restricted, the delivery of bikes was also sporadic.

Development of the R75 was concluded in February 1941, with the first motorcycle leaving the production line in July 1941. After the construction of 6,000 examples, more space was required in Munich for aircraft engine manufacture and production moved to Eisenach from July 1942 (against Popp's recommendation). The GBK (Bike Select Committee) also decided that the Zündapp KS750 was a better machine than the R75, and by August 1942, instructed BMW to cease R75 manufacture in favour of the KS750. The R75 was proving too expensive to produce, and the front forks too weak for the heavy loads. The GBK also requested the R75 to be equipped with the KS750 parallelogram fork, but because BMW still had 5,000 telescopic forks in stock, this did not happen. Not surprisingly perhaps, but BMW was also reluctant to embark on the production of a competitor's motorcycle on the command of the Wehrmacht.

The R75 earned a reprieve and production continued at Eisenach, but this stalled initially because German workers were reluctant to move from Munich, as they would lose their reserved occupation status and were likely to be called up for military service. BMW therefore requested foreign workers, and around 1,000 Russian prisoners of war were trained to manufacture R12s and R75s. As a result, R75 production numbers were considerably below expectations, with around 2,000 less than was anticipated by the end of 1942.

Circumstances changed during 1943. A shortage of raw materials saw the aluminium castings sourced from outside with a resulting loss in quality. Combined with foreign workers who lacked training and motivation, production costs escalated as it now took a further two hours to repair defects on each machine after the first test drive. At a cost of 2,000 Reich marks each, the Wehrmacht placed their final order of 2,000 machines during 1943 and it was inevitable that R75

production would then cease. Not only was the R75 complicated, expensive, and considered inferior to the KS750, but also the army now preferred the cheaper, mass-produced four-wheeled VW Kübelwagen. Despite some export orders (to unspecified armies), the initial target was to cease R75 production in May 1944 (with chassis number 770200) to release the workers for aircraft manufacture. This was later amended to the end of December 1944 when it became clear this goal couldn't be met.

By the end of March 1944, 17,635 machines had been delivered, but air raids in July began to interrupt production. After three more raids production ended on 18 October 1944, and when US troops occupied the factory in April 1945, 60 per cent of the buildings were destroyed. Eisenach subsequently became part of East Germany and 98 R75s were produced out of spare parts during 1946 and delivered to Russia. Although the EMW factory began developing an updated R75 military motorcycle during 1952, this never reached the production stage.

At around the same time as the air raids on Eisenach, the Allies began bombing BMW in Bavaria in earnest. There were now two plants, one in Allach and are in Milbertshofen at Oberwiesenfeld. Allach was to the north west of Munich, close to the concentration camp at Dachau, and was largely spared allied bombing, but the RAF bombed Milbertshofen as early as September 1940. There was another RAF raid in March 1943, but the most damage was sustained when the US Air Force began their bombing on 9 June

Although around 18,000 R75s were produced, most were destroyed during the war and only a few have survived. This is an R75 in original Wehrmacht trim at the 1976 Lion Rally in Zolder, Belgium. *Damien Cook*

1944. After eight air offensives, through until 22 September 1944, more than half of the Milbertshofen works was destroyed. Although stripped and looted, the undamaged Allach works became a US Army supply and transport depot.

Earlier, in June 1942, Popp resigned as chairman of the board after differences of opinion with the Reich's Aviation Ministry, but remained a member of the Supervisory Board of BMW AG until 8 May 1945. Popp was always wary of the Nazis, maintaining his loyalty to BMW rather than the Third Reich, something he reiterated in the denazification court of 1947. On 11 April 1945, Hitler issued his 'scorched earth' policy that required the destruction of all military assets, including BMW, but Albert Speer thwarted this on the grounds that, 'we should not destroy what generations have built up before us.' This still didn't save BMW as the company was on the

Allies' blacklist, probably due to its development of the 003 jet engine. On 1 October 1945, the order came for the levelling of BMW factories 1 and 2, and it looked unlikely that the company would survive.

This was where Kurt Donath, Milbersthofen works manager since 1942, and non-Nazi party member appointed by the Allies in 1945, intervened. He managed to save the works and initiate the manufacture of saucepans, agricultural machinery, and bicycles. But a stroke of luck enabled BMW to rebuild as a motorcycle manufacturer. Currency reform and a relaxation of economic restraints during 1948 meant confiscated assets suddenly became 'free', and it was discovered that the Reich owed BMW 63.5 million marks from Deutsche Bank cheques dating back to 28 April 1945. Donath now had the money to get the company rolling again, and just as BMW did after the First World War, he decided to build motorcycles.

Many examples of the R25/3 of
1954 came factory-fitted with a
sidecar. *BMW Mobile Tradition*

3 SINGLE CYLINDERS

As the R32 twin was sold as a premium motorcycle, production levels were necessarily low and the profitability in difficult times was dubious. BMW's management though, wanted to expand aircraft engine production and develop a car, and they insisted the motorcycle operation should become profitable. This required building a lower priced model to supplement the R32, opening the door to a wider range of buyers. Development of a single-cylinder model began in April 1924, and the R39 made its first appearance at the Berlin Motor Show at the end of the year. The R39 was the first of a range of single-cylinder models that lasted through until 1966. However, despite often displaying an advanced specification, the singles always only provided adequate performance and have never garnered the following of the twins, particularly outside Germany. Within Germany though, the singles often outsold the more expensive boxers, especially during the 1930s and 1950s. In the two decades, from 1948 through until 1969, 62 per cent of BMW motorcycles produced were singles.

THE R39

Tyre supply problems saw production of the R39 delayed until September 1925, and while it was intended as a budget model, the specification was surprisingly high. Like the boxer twins, the 250cc engine (designation M40a) was mounted longitudinally in the frame, but with a vertical cylinder, and shaft drive. The three-speed gearbox was bolted directly to the crankcase and driven through a single-plate dry clutch mounted to the flywheel. The alloy overhead-valve cylinder head came from the sporting R37, and provided the R39 with a conservative 6.5bhp at 4,000rpm.

The R39 frame was a twin-tube design like the R32, but the rear brake was an external shoe acting on a drum on the driveshaft, rather than the earlier wedge-shaped brake block. With a Bosch magneto-generator lighting set and speedometer the R39 cost 2,150 marks, almost as much as an R32. The performance for a 250cc was outstanding, with a top speed of around 100km/h (62mph) – more than the R32. Sepp Stelzer rode an R39 to victory in the 1925 250cc German road racing championships, but while sales were initially strong they dwindled following reports of problems

The first BMW single was the R39, and while the performance was excellent for a 250 single it was too expensive. Few were manufactured and only a handful has survived. This example is in a private collection in Bavaria. *Ian Falloon*

with cylinder bore wear and excessive oil consumption. Production ended in 1926 but the R39 was offered well into 1927.

THE R2

The failure of the R39 deterred BMW from producing another single until 1931. By this time however, the economic circumstances and registration requirements in Germany were quite different. After 1 April 1928 the registration rules were amended so that motorcycles under 200cc could be used without road tax and ridden without a licence. This saw DKW produce large numbers of mass-produced two-strokes, and while BMW wasn't initially interested in competing in this market, they were forced into it following the deteriorating economic climate after the Wall Street Crash of October 1929.

Unlike the R39, BMW managed to find the right formula with the R2. It was no easy feat, trying to combine traditional BMW four-stroke quality and a price people could afford, but BMW managed to do this. Also retaining shaft drive, the 198cc R2 engine (M67a) was mounted longitudinally in the frame, with a three-speed gearbox bolted behind. The overhead valves were exposed and the crankcase a one-piece tunnel design, a feature of all later air-cooled BMWs. Ignition was by battery and coil and the power was an acceptable 6bhp which was enough to propel the rather heavy (130 kg/287lb) R2 to around 95km/h (59mph), with excellent fuel economy.

The F57 pressed-steel frame and nine-plate cantilever forks followed the lines of the R11 and R16, but the engine was offset to the right to allow for a direct driveline in top gear, from the crankshaft, to minimise power loss. An improvement was the internally expanding rear drum brake instead of the earlier driveshaft brake. The R2 was designed with practicality and ease of ownership in mind, even including a front stand to assist wheel removal.

New registration requirements led to the 200cc R2 for 1931. *BMW Mobile Tradition*

Although it sold for 975 marks, three times the cheapest DKW, the R2 was immediately successful.

SERIES 2A & 2/33

For 1932, the R2 engine (M67 S II) received enclosed valve gear, a SUM type K5/250 carburettor, and a new gearlever. There was a second Series 2 for 1933, 80 examples having an Amal carburettor, but the main changes were to the F67 S II 33 chassis which, from June 1933, incorporated a friction damper with scissors on the front fork.

SERIES 3

All M67 S III engines for 1934 received the Fischer-made Amal push-in carburettor and a revised camshaft to increase the power to 8bhp. The 6-volt, 30-watt generator on the left side of the engine was now under an aluminium cover. The wheelbase was also reduced slightly, to 1,303mm.

SERIES 4

For 1935 (M67 S IV engine, F67 S IV chassis), there was a reshaped, smaller, and longer fuel tank, and a new type of Bosch headlamp. A rubber saddle also replaced the leatherette type.

SERIES 5

The final R2 of 1936 (M67 S V engine, F67 S V chassis) featured an Amal type 74/412S carburettor, a revised driveshaft, and a wider rear wheel mudguard and numberplate. With over 15,000 produced, the R2 was one of BMW's motorcycle production mainstays between 1931 and 1936. The R2 may now seem an insignificant model, but it was pivotal to the survival of the company in the early 1930s.

The R2 Series 2a of 1932 featured
enclosed valve gear. *Brian J. Nelson*

THE R4 & R3

Following the demise of the 500cc R52 and R57 in
1930 there was a gulf in the line-up between the 200cc
R2 and 750cc R11 and R16. Rather than build another
expensive twin, BMW opted for expediency, creating
the 400cc R4 single out of the R2 for 1932. The 78 x
84mm engine (M69 S1) was based on the R2, but
always featured enclosed valves. The kickstart was on
the right and the engine produced 12bhp, enough for
100km/h.

 Although the chassis (F69 S1) was essentially
the same as the R2, the pressed-steel forks featured
additional strengthening steel strips, the front
mudguard was valanced, and the tyres were a
slightly larger section. From July 1932, the forks
incorporated a friction damper, but this was
largely ineffective.

SERIES 2

For 1934, the M69 S2 engine received a four-speed
gearbox, there was a new gear lever and starter (no
longer sideways), and a copper-wool air filter for the
Sum CK 3/500 Fr carburettor. The F69 S2 chassis
incorporated some styling revisions, an extender spring
saddle, and new rubber-covered footrests in place of
the alloy boards.

SERIES 3

A new cylinder head for the M69 S3 of 1934 saw the
power up to 14hp, with the generator on the left now
under a cover. The toolbox was also incorporated in
the gearbox casting. Chassis changes (F69 S3) included
a larger fuel tank (12 litres), with the gearshift now in
a gate near the fuel tank, as on the R11 from 1933.

A 400cc single, the R4, was available from 1932, lasting until this final Series 5 version appeared in 1936. *Brian J. Nelson*

SERIES 4

The toolbox was now integrated into the crankcase (M69 S4) for 1935, with the generator on top of the crankcase driven by a V belt from the crankshaft, and the battery in a separate box near the gearbox. The F69 S4 chassis received a new headlamp and the forks had twin friction dampers.

SERIES 5

For 1936–37 (M69 S5), the gearcase was changed, with new gears, and the oil riser moved. The chassis (F69 S5) was much as before, and this final series was produced in relatively large numbers for military use.

As the R4 wasn't able to compete with the more powerful ohv 500s it was an alternative to the more mundane side-valve models. The price of 1,150 marks also deterred it from younger buyers but it soon earned a reputation for ruggedness and reliability that saw it become the standard training and dispatch model for the military and police. It was also used in trials events and was offered for a while as an off-road sports model, albeit without any modifications from standard.

Joining the range of singles for 1936 was a small-bore version of the R4, the R3. The long-stroke 305cc single (M203/1) produced 11bhp, but not having the performance of the R4 or the tax and licence advantages of the R2, it was discontinued after only one year.

THE R35, R20, R23

The final pressed-steel frame single was the R35, replacing the R4 during 1937. A reduction in the cylinder bore to 72mm gave 342cc, but the power

output of the M69a-derived engine (M235/1), with a SUM CK 9/22 carburettor, remained at 14bhp. New for the R35 was a set of telescopic forks, although these were rather rudimentary in design and didn't incorporate the hydraulic damping of the R5. The R35 was very much an anachronism at this time of technical innovation at BMW, but despite this, it was very popular with the German military which bought it in large numbers, just as they had the R4. Although the main production ended in 1940, more were produced from 1947 at the Eisenach plant, now in East Germany, to use up stocks of parts. The EMW (Eisenacher Motorenwerke) R35 and R35-3 (with plunger rear suspension) remained in production through until 1955. Of the 80,000 examples completed, only a handful ever made it to the West.

During 1937, BMW also envisaged 500cc and 350cc singles, the R36 (M236/1). This design had both inlet and exhaust ports facing the rear of the engine, and the engine was rubber mounted. Following delays from the military, development was suspended in mid-1939.

More modern than the R35 was the R2 replacement, the R20 with its tubular steel frame, also introduced for 1937. The ohv engine (M220/1) was an all-new design, with dimensions of 60 x 68mm, producing 8bhp. There was a foot-change for the three-speed gearbox and the generator was now driven from the crankshaft and positioned in front of the timing cover. The frame of the R20 consisted of butted-end tubes bolted together, and the telescopic forks were the undamped R35 type. After June 1938, new traffic regulations in Germany no longer exempted 200cc machines and there was now a restricted licence for motorcycles up to 250cc. BMW then responded by creating the 250cc R23 by boring the R20 engine to 68mm (M223/1). The power went up to 10hp and apart from a toolbox which was now incorporated inside the fuel tank, the R23 was identical to the R20. Large numbers of both the R20 and R23 were produced, but the outbreak of the Second World War saw the demise of production.

THE R24, R25, R25/2 & R25/3

When BMW decided to recommence motorcycle production after the Second World War, Allied requirements restricted the displacement to 250cc. Also, all the drawings for all BMW products were requisitioned by the Americans so a pre-war R23 was stripped down and minutely measured. The plans were complete by Summer 1947, but a complete machine took a further year to materialise. When the R24 was first shown at the Geneva Show in March 1948, most of the basic components were missing (but represented by wooden mock-ups), although at the Export Fair in Hanover in May, it was only minus a gearbox, gearwheels and crankshaft. There was an overwhelming response to the R24, with 2,500 advance orders, but material shortages delayed commencement of production until December 1948.

Although largely based on the R23, the R24 engine (M224/1) featured a new cylinder head, strongly influenced by the design of the wartime R75. The valve angle was changed, the rockers worked in bolted-on pillars rather than cast bosses, and the pushrods were led in through tunnels in the cylinder head. Like the R75, the valve covers were in two pieces, held by a clamp with a single bolt. The compression ratio was increased slightly, to 6.75:1, and with a Bing AJ1/22/140b 22mm carburettor the power was up to 12bhp at 5,600rpm. The electrical system was from Noris in Nuremberg, with the battery ignition incorporating a centrifugal advance. Also new was a four-speed gearbox. The chassis was similar to the R23, with a bolted, rigid tubular steel frame and telescopic forks.

With a chromed fishtail exhaust and trim embellishments on the mudguards, the finish and appearance were of high quality. The R24 was also the most expensive German motorcycle, but undoubtedly Schorsch Meier's exploits on the 500 Kompressor boosted sales, and 9,400 were sold in 1949. Production continued through until 1950 when the R25 replaced it.

There were only a few developments for the R25

During the early 1950s, singles like
this R25/3 were extremely popular
in Germany with production
considerably outstripping twins.
Ian Falloon

Opposite: Earles forks and a rear
swingarm graced the R26, which
was the most popular BMW
motorcycle in the late 1950s.
BMW Mobile Tradition

engine (M224/2), including a stronger crankshaft, 2mm larger inlet valve, and a Bing 1/22/28 24mm carburettor. More significant was the welded tubular steel frame that was suitable for sidecar attachment, and plunger rear suspension for the first time on a BMW single. Also setting the R25 apart was the deeply valanced front mudguard.

The similar R25/2 appeared during 1951, the engine (M224/3) reverting to the earlier R24 specification. Either a Bing 1/22/44 or SAWE K22F 22mm carburettor was fitted and there were only detail changes to the cycle parts (designation 225/1). For the R25/3 of 1953 the engine (M224/4) there was a higher compression ratio (7:1), with improved intake and exhaust system, and a 24mm Bing 1/24/42 or SAWE K24F carburettor for a power increase to 13bhp at 5,800rpm. The chassis (225/3) included hydraulically damped forks, full hub width brakes, 18in alloy

wheels, and a larger fuel tank that incorporated the induction pipe. During the early 1950s, the R25 singles proved extremely popular in Germany. This was a period where cheap transportation was in demand, and while the BMW 250s only provided barely adequate performance, they were quality machines, built to last. However, circumstances had changed by 1955. The domestic market had reached saturation and a more modern motorcycle was required for export.

THE R26 & R27

Following on from the R50 and R69 of 1955 (see Chapter 4), the 250cc single also received Earles-forks and a rear swingarm for 1956. The new model was the R26, and while motorcycle sales were generally extremely depressed during the latter half of the 1950s, the R26 proved surprisingly popular, particularly for

The final shaft-drive BMW single was the R27, now with a rubber-mounted engine to isolate the vibration. *BMW Mobile Tradition*

BMW R 27

BMW R 27
Antrieb:
Motorleistung 18 PS, bei Drehzahl 7400 U/min
Bohrung und Hub 68 mm/68 mm, Verdichtungsverhältnis 8,2:1
Kraftübertragung:
Getriebeübersetzung I 5,33 II 3,02 III 2,04 IV 1,54
Hinterachsübersetzung solo 4,5:1, wahlweise 4,16:1, mit Beiwagen 5,2:1
Bremsen:
Leichtmetall-Vollnabenbremsen, vorn Duplex-, hinten Simplex-Bremse
Abmessungen: größte Breite 660 mm, größte Länge 2090 mm, Sattelhöhe 770 mm
Gewichte: fahrfertig solo 162 kg, zulässiges Gesamtgewicht solo 325 kg
Lichtanlage: 6 Volt/60 – 90 Watt
Tankinhalt: 15 Liter
Kraftstoff-Normverbrauch: 3,9 l/100 km
Höchste Geschwindigkeit: solo 130 km/h

BMW R 50
Antrieb:
Motorleistung 26 PS bei Drehzahl 5800 U/min (Sonderwunsch: mech. Drehzahlmesser-antrieb), Hubraum 494 ccm, höchstes Drehmoment 3,5 mkg
Bohrung und Hub 68 mm/68 mm, Verdichtungsverhältnis 7,5 : 1
Kraftübertragung:
Getriebeübersetzung I 4,171 II 2,725 III 1,938 IV 1,54
m. Beiwagen 5,33 3,02 2,04 1,54
Hinterachsübersetzung solo 3,13:1, wahlweise 3,58:1, mit Beiwagen 4,33:1
Bereifung: solo 3.50 – 18, mit Beiwagen hinten 4.00 – 18
Bremsen: Leichtmetall-Vollnabenbremsen, vorn Duplex-, hinten Simplex-Bremse
Abmessungen: größte Breite 660 mm, größte Länge 2125 mm, Sattelhöhe 725 mm
Gewichte: fahrfertig solo 198 kg, zulässiges Gesamtgewicht solo 360 kg
Lichtanlage: serienmäßig 6 V/60–90 W. Behörden und Polizei 12 V/100–150 W
Tankinhalt: 17 Liter
Kraftstoff-Normverbrauch: 5,1 l/100 km
Höchste Geschwindigkeit: solo 140 km/h

BMW R 60
Antrieb:
Motorleistung 30 PS bei Drehza
antrieb), Hubraum 594 ccm, höc
Bohrung und Hub 72 mm/73 m
Kraftübertragung:
m. Beiwag
Hinterachsübersetzung solo
Bereifung: solo 3.50 – 18, mit B
Bremsen: Leichtmetall-Vollnabe
Abmessungen: größte Breite 66
Gewichte: fahrfertig solo 198 k
Lichtanlage: serienmäßig 6 Vo
Tankinhalt: 17 Liter
Kraftstoff-Normverbrauch:
Höchste Geschwindigkeit: solo

Änderungen von Konstruktion und Ausstattung im Interesse der technischen Weiterentwicklung vorbehalten · Bayerische Motoren Werke AG, München   Printed in W

export markets in Third World countries. More than half the motorcycle production between 1956 and 1960 comprised the R26.

The R26 chassis (226/1) was very similar in layout to the R50, providing improved handling and comfort over the old plunger-frame R25/3. The weight was also increased slightly, so the engine (M224/5) was uprated. No longer painted black, the cylinder head had larger cooling fins, a higher compression ratio (7.5:1), and a larger, Bing type 1/26/46 carburettor to produce 15bhp at 6,400rpm. There was also an aluminium con-rod, with the big-end running directly on the rod journal. Tests with the aluminium con-rod had begun back in 1951, but later, BMW returned to the steel con-rod and roller bearing big-end. During 1956, an engine with an overhead camshaft driven by a vertical shaft was considered, with another designed by Ludwig Apfelbeck in 1959. Both these designs were deemed too expensive.

During the restructuring of BMW in 1960, the R26 was developed, becoming the R27. Although essentially similar, the engine (226/2) now produced 18bhp, at a considerably higher 7,400rpm. The compression ratio was increased to 8.2:1, and there was a new camshaft. The contact breaker was positioned on the front of the crankshaft, and there was a spring-loaded tensioner for the timing chain. The most significant development though was the engine which was now rubber mounted to quell the increased vibration. The engine and gearbox were supported by rubber mounts in four locations in the duplex frame, along with a rubber cylinder head bracket and two fore and aft rubbers on the front and rear to limit longitudinal movement. In all other respects though the chassis (226/2) was identical to the R26.

The R27 was intended to be the biggest seller in the revamped motorcycle range of the early 1960s, but it was too expensive and still only offered barely adequate performance. After production ended in 1966, efforts were made to secure an order from the German Federal Armed Forces for a revamped model, the R28. During 1967, various prototypes with a new duplex frame and a variety of forks (including telescopic) were tested, but BMW's bid failed and the contract was awarded to Hercules.

Production of BMW singles numbered more than 230,000 over their 41-year lifespan, and they served a useful purpose in providing reliable and inexpensive transport. However, by the end of the 1960s, motorcycling was increasingly becoming a leisure activity, and buyers demanded more performance. It looked as if the days of single-cylinder BMWs were over, but times change, and 27 years later it was resurrected in the form of the F650 which soon followed the R25 and R26 in eclipsing the twins to become the best-selling model in the BMW motorcycle line-up.

nderwunsch: mech. Drehzahlmesser-
t 4,2 mkg
häitnis 7,5:1

III 1,938 IV 1,54
    2,04   1,54
ntrieb 3,86:1
00 – 18
plex-, hinten Simplex-Bremse
ge 2125 mm, Sattelhöhe 725 mm
mtgewicht solo 360 kg
örden und Polizei 12 V/100 – 150 W

Beiwagen 6,5 I/100 km
agen ca. 110 km/h

14 196 10 X 66

BMW R 69 S
Antrieb:
Motorleistung 42 PS bei Drehzahl 7000 U/min (Sonderwunsch: mech. Drehzahlmesser-
antrieb), Hubraum 594 ccm, höchstes Drehmoment 4,45 mkg
Bohrung und Hub 72 mm/73 mm, Verdichtungsverhältnis 9,5 : 1
Kraftübertragung:
Getriebeübersetzungen I 4,171 II 2,725 III 1,938 IV 1,54
Hinterachsübersetzung 3,13 : 1
Bereifung: 3.50 S 18
Bremsen:
Leichtmetall-Vollnabenbremsen, vorn Duplex-, hinten Simplex-Bremse
Abmessungen: größte Breite 722 mm, größte Länge 2125 mm, Sattelhöhe 725 mm
Gewichte: fahrfertig 202 kg, zulässiges Gesamtgewicht solo 360 kg
Lichtanlage: serienmäßig 6 V/60-90 W. Behörden und Polizei 12 V/100-150 W
Tankinhalt: 17 Liter
Kraftstoff-Normverbrauch: 5,3 I/100 km
Höchste Geschwindigkeit: solo 175 km/h

Klaus Enders and Ralf Engelhardt on their way to winning the 1972 Dutch TT, and the Sidecar World Championship. *Mick Woollett*

4 POST-WAR TWINS

During the summer of 1949, the Allied force's restriction on motorcycle displacement was lifted for German motorcycle manufacturers, opening the door for the reintroduction of the 500cc boxer twin. As there was still little money for development, the 500 twin was initially based on a pre-war design, in this case the 1938 R51. BMW was fortunate that the R51 was an advanced design for its day, and although its foundations went back to the even earlier R5, the R51/2 proved immediately popular when it was released in 1950.

THE R51/2

Apart from the split valve covers (similar to those of the R75), and Bing 1/22/39 and 1/22/40 carburettors, the engine of the R51/2 (M254/3) was virtually identical to the pre-war R51. There were new cylinder heads with coil valve springs, but the roller-less chain-driven twin camshaft set-up was the same. Ignition was still by battery and coil, and the R51/2 retained the exposed generator with its distinctive finned clamp on top of the engine. There was now a coil spring damper on the gearbox mainshaft, and a revised lubrication system with pressurised oil to the camshaft bearings. Although the frame (251/2) was the same as the final R51 of 1941, with two additional strengthening tubes, the telescopic forks gained two-way damping. Finally, the old-style inverted pivot control levers gave way to the more usual pivot type. The R51/2 certainly re-established BMW as a prominent motorcycle manufacturer, but as a pre-war design, it was always only going to be a stopgap. After only a year, BMW released the R51/3, with a new engine.

THE R51/3, R67, R67/2 & R67/3

Apart from slightly different pinstriping on the front mudguard, the R51/3 of 1951 looked almost identical to the R51/2. Despite the power output of 24hp remaining the same, inside the redesigned engine cases was a completely new engine (M252/1) that would power all twins through until 1969. The crankshaft was still a built-up type, with two ball bearings at the

The first post-war twin was the
R51/2, which was essentially the
same as the pre-war R51 but
the split valve covers were new.
Ian Falloon

Opposite top: Although it looked
similar to the R51/2, the engine
was new for the R51/3 of 1951.
BMW Mobile Tradition

Opposite bottom: The 600cc R67
was sold primarily for sidecar use,
but was never especially popular.
Ian Falloon

front and a ball bearing at the rear. The con-rods also
ran on roller bearings. Replacing the twin camshafts
and long timing chain that was always susceptible to
whipping, was a single camshaft above the crankshaft.
Driven by helical gears, this allowed for a much
narrower crankcase, with a Noris magneto driven from
the crankshaft and a contact breaker and automatic
advance from the front of the camshaft. A third gear
drove the oil pump and as everything was encased in
smooth new covers the engine appeared to be much
more modern. Gone too were the two-piece valve
covers, and there were new cylinders and heads. The
34mm and 32mm valves retained the 80° included
valve angle. The paper-element air filter was also now
mounted in a special casing above the gearbox.

The engine may have looked more up-to-date, but
the chassis (251/3) was almost identical to that of the
R51/2. For 1952 (251/4) there was a duplex 200mm

front brake, and rubber gaiters for the front forks,
while for 1954, the front brake was a full-width duplex
type. Also, for 1954 there were alloy wheel rims, a
fully covered air filter, and new mufflers.

Alongside the R51/3 for 1951 was the similar R67. A
bore and stroke of 72 x 73mm provided 594cc, but the
R67 (M267/1) was intended mainly as a sidecar machine
rather than a sportster like the pre-war R66. A power
output of only 26bhp provided sedate performance, and
for 1952 and the R67/2 (M267/2) this was increased
marginally to 28bhp. The R67 chassis was identical to
that of the R51/3, and the R67/2 received the 1952 R51/3
improvements. The R67/3 of 1955–56 was the final
BMW twin to retain the plunger rear end, and with its
4.00 x 18in rear tyre was even more suitable for
sidecar haulage. However, the new Earles fork models
were more appropriate for sidecar attachment and the
R67/3 remained a relatively unpopular budget model.

Alfred Böning, one of the most
important BMW motorcycle
engineers for more than 40
years, pictured in 1960.
BMW Mobile Tradition

Opposite: One of the most coveted
post-war BMW motorcycles is the
high performance, and rare, R68.
Ian Falloon

ALFRED BÖNING

Involved in the design of nearly every BMW motorcycle for over
four decades, Alfred Böning was one of BMW's most significant
engineers. Böning was born of German parents in Salerno,
Southern Italy, in 1907, moving to Esslingen in Southern Germany
in 1917. He graduated from engineering school in 1930, and after
working at NSU, Böning went to BMW in November 1931. One of
the few senior engineers to remain with BMW immediately after
the war, Böning ended up staying with the company until 1972.
Initially employed as a motorcycle engineer, Böning designed
frames, gearboxes and rear-wheel drives for motorcycles, and was
responsible for the R12, R5, and R75. During 1943, he became
manager for aircraft engines, including supervising the design of
rocket and jet engines.

In May 1945, Kurt Donath, director of the Munich plant,
instructed Böning to rebuild the development department away
from the factory, which was still in ruins. He was subsequently
responsible for all the first post-war motorcycles, and also the
Isetta 600 and 700 cars. After handing over the design
department to von Falkenhausen in 1960, Böning remained a
senior engineer, and in 1968, became director of the development
of car and motorcycle chassis and transmissions.

THE R68

As motorcycle production continued to climb, from
9,400 in 1949 to 25,101 in 1951, BMW needed a
sporting flagship to compete with the new 100mph
British Triumph and BSA twins. Their response was the
R68, first displayed at the Frankfurt Motorcycle Show
at the end of 1951. Here was the first real sporting
BMW motorcycle since the R66, and it has rightly
earned a place as one of the most desirable post-war
production models. Not only did the R68 look far
more purposeful than the R67, it also provided
improved performance, with a claimed top speed of
100mph (161km/h).

The uprated 594cc R68 engine (designation 268/1)
included 8:1 pistons, 38mm and 34mm valves, a fiercer
camshaft, rockers pivoting on needle rollers under the
new twin-rib valve covers, and a barrel-shaped roller
bearing for the rear of the crankshaft. With Bing 1/26/9
and 1/26/10 carburettors, the power was increased to
35bhp at 7,000rpm. The initial show bike featured an
upswept two-into-one exhaust system resembling that
of the 1951 Varese International Six-Day Trial factory
R51/3 racers of Meier, Zeller, and Kraus. This exhaust
system didn't make it to the 1952 production R68 that
wore standard fishtail exhausts, but the 2-1 remained
an optional accessory. Also setting the R68 apart were
finned exhaust clamps.

Although the R68 chassis (251/4) was essentially
identical to the R67/2, it was distinguished by several
features. The front mudguard was narrower, with a steel
brace, and an optional sprung pillion pad was available,
although this was primarily to allow the rider to adopt
a more prone riding position. The claimed top speed
was 160km/h (100mph), and the brakes were the same
200mm duplex of the 1952 R51/3.

There were few changes for 1953. By late 1952
rubber gaiters appeared on the front forks, and the
mufflers were now non-finned. After July 1953, a
sidecar mount was provided on the frame, and later
that year the 100,000th BMW motorcycle rolled off
the production line. There were more developments for
1954. Along with light alloy wheel rims, the front
brake was full width and there was a larger headlamp.
At a price of DM3,950, the R68 continued a BMW
tradition that made it available only to a fortunate few,
and with only 1,452 produced, the R68 remains one of
the rarest post-war BMW motorcycles.

THE RS54

As soon as the FIM allowed Germany to compete in the World Championship in 1951, Leonhard Ischinger created a new racing engine (M253a) strongly based on the pre-war Kompressor, but without the supercharger and with twin carburettors. This retained telescopic forks and the plunger rear end, but with only 43bhp at 8,000rpm was never going to provide international glory. Although Walter Zeller won the 1951 German Championship, BMW decided to produce a motorcycle suitable to promote racing and provide German riders with an affordable competitive machine. The M253b followed during 1952, with the M253c for 1953. After experimenting with square, 68 x 68mm dimensions (as

on the production R51/3), and a narrower 60° included valve angle, the M253c eventually retained many of the basic features of the Kompressor. This included the 66 x 72mm bore and stroke, a wide included valve angle of 82°, the close-coupled twin overhead camshafts driven by bevel gears, and only two main crankshaft bearings. As the cylinders were offset, the bevel drive lined up with the exhaust camshaft on the right and the inlet on the left, the second camshaft coupled directly to the driven shafts. The cams were too close together to actuate the valves directly, so rockers were used, resulting in the wide valve angle. The power was around 50bhp at 8,500rpm.

Only a few RS54 500cc racers were
produced, and they were extremely
expensive and not very competitive
in standard trim. *Ian Falloon*

A small M253 production series was also offered
during 1953. These were the RS54 (Rennsport), and
were so expensive to produce that the factory lost
almost as much on each machine as it made. At around
DM6,000 each, only 24 were manufactured. The
catalogue engine had a low, 8:1 compression ratio and
with two Fischer-Amal 30mm carburettors produced
45bhp at 8,000rpm. The RS54 also had a new frame,
with an oval section top tube, duplex loops, and a
pivoted fork at both ends. At the front was an Earles-
pattern leading link type, and for the first time the
driveshaft was enclosed in the right fork arm. There
was a twin leading shoe front brake, and because the
RS54 weighed only 130kg (287lb), the performance
was surprisingly good.

During 1953, there was experimentation with fuel
injection on the works M253, the first stage spraying
through the sides of the inlet tracts, with the second
featuring axial sprays into the air trumpets. This
featured on Walter Zeller's 253 at the Isle of Man in
1953. Zeller crashed on the second lap but it was
evident the 253 wasn't yet competitive in the solo
500cc category. It was another story altogether in
sidecar racing, and Wilhelm Noll and Fritz Cron
showed that even by simply attaching a Steib sidecar to
the RS54 solo machine it was competitive. Already the
engine layout was proving ideal for sidecar racing, the
low, wide engine facilitating safe drifting and the shaft
drive an asset rather than a hindrance. Encouraged by
the performance of the works 253 racers, during 1953

Walter Zeller's factory RS racers generally featured fuel injection, and took him to second place in the 1956 500cc World Championship. *Ian Falloon*

the plans were laid for a 500cc four-cylinder boxer engine with direct fuel injection. Titled the Haufenmotor, or 'heap engine', the rear cylinders were angled 20° upwards for better cooling. There were also plans for an even more complicated twin-crankshaft engine, but neither was built.

On the works M253, the fuel injection was further developed for 1954, with fuel now pumped directly into the cylinder heads through nozzles opposite the spark plugs. Valve sizes were 40mm and 36mm, and a high-domed four-ring piston provided 10.2:1 compression. As there were still only two main bearings the safe maximum revs were 9,000rpm. Zeller won the German Championship but didn't achieve any spectacular international results. Noll and Cron's

works sidecar racer now featured streamlining, and with three victories, they easily won the Sidecar World Championship.

In October 1954, Alex von Falkenhausen returned to head the competition department and there was more development for 1955. Carburation was either by Dell'Orto carburettors or fuel injection, and Zeller won the German Championship, finished second at the German Grand Prix, and 10th in the 500cc World Championship. One of von Falkenhausen's developments was to patent a pivoted rear-wheel drive housing by Helmut Werner Bönsch. Zeller's machine from 1955 thus featured a modified drive system with external drive shaft, and his performances during 1955 prompted the factory to further develop the machine

Fritz Hillebrand and Manfred
Grünwald at the Isle of Man, 1957
where they won the TT and went
on to take the Sidecar World
Championship. *Mick Woollett*

Opposite top: Geoff Duke (20) and
Dickie Dale (standing on the right)
before the start of the 1958 Isle of
Man Senior TT. *Mick Woollett*

Opposite bottom: Dickie Dale on
the RS at the Austrian Grand Prix
at Salzburg, 1958. *Mick Woollett*

and contest the entire World Championship the
following year. Willy Faust and Karl Remmert headed
a display of complete BMW dominance in the sidecar
class during 1955 with their BMW/Steib.

For 1956, the works racers (M253f) had a bore and
stroke of 70 x 64mm, and a five-speed gearbox.
Depending on the circuit, Dell'Orto carburettors or
fuel injection were used, the engine producing 58bhp
on carburettors and 61bhp with injection, although
low-speed response was inferior. Handicapped by a
decision not to run a fairing due to the high winds,
Zeller managed fourth at the Isle of Man, but followed
this with second places in Holland and Belgium. Zeller
failed to win a Grand Prix but finished second to John
Surtees in the 500cc World Championship. That year,
former World Champion Fergus Anderson also rode
the RS, but was killed on one at Floreffe in Belgium.
BMW again dominated sidecar racing in 1956, the
fully streamlined 'Barquette' RS of Noll and Cron, now

with an integral BMW sidecar and 16in wheels, taking
out the World Championship.

With the Gilera fours back to full strength, the
BMWs struggled in the 1957 500cc World
Championship, although Ernst Hiller and Zeller
managed some respectable results. Fritz Hillebrand and
Manfred Grünwald continued BMW's dominance of
the sidecar championship. Zeller retired at the end of
1957 and Geoff Duke and Dickie Dale were provided
with works machines for 1958, but without official
support. Dale had some reasonable results, finishing
second in the Swedish Grand Prix and fourth overall in
the World Championship, while Duke never managed
to come to terms with the idiosyncratic BMW.

Later in 1958, a 250cc boxer twin (M207) was
constructed which was a resurrection of a 1933 design
for a prototype R7. With a vertically split magnesium
crankcase it featured a centre crankshaft bearing and
revved safely to 11,500rpm. A desmodromic cylinder

Alexander von Falkenhausen astride an R90S in 1975. *BMW Mobile Tradition*

Opposite: Walter Schneider at the 1957 Belgian Grand Prix. With Hans Strauss, he won the race, and went on to win the Sidecar World Championships in 1958 and 1959. *Mick Woollett*

ALEXANDER VON FALKENHAUSEN

Alexander von Falkenhausen masterminded BMW's domination of sidecar racing until 1964, but his influence on the general technical development of BMW motorcycles was even wider ranging. Born in 1907, von Falkenhausen graduated from the Munich College of Advanced Technology in 1934 and went directly to BMW as a motorcycle frame and suspension designer, with a contract as an off-road competition rider following his success during 1933 on a British Calthorpe motorcycle. By 1936, he was in charge of motorcycle gearbox design, and in 1937 became head of motorcycle experimental development. From 1943 until 1945, von Falkenhausen was manager of engine development. Before the war, he was a successful competitor, winning three gold medals in the 1936 ISDT before changing to cars. With his private BMW 315, von Falkenhausen took 35 victories, switching to a BMW 328 that took him to more than 100 race wins through until 1953.

After the war, von Falkenhausen established his own company AFM (Alex. von Falkenhausen, Munich), building sports and racing cars. He won the 1948 German Sports Car Championship in a 1500cc AFM, but also continued to race his BMW 328. Soon after rejoining BMW in 1954, he oversaw the preparation of Wilhelm Noll's RS54 speed record sidecar. With a fully streamlined machine, Noll set a new absolute sidecar world speed record of 280.20km/h (174.12mph) in 1955. In 1966, von Falkenhausen took to the wheel himself, setting two world records in a BMW Formula 2 racing car over a quarter-mile and 500m. From 1957 he was again in charge of BMW engine development, until retiring in 1975.

head was also produced for the 500cc (M253/d), but both projects were abandoned due to financial considerations. Dale rode the RS again during 1959, finishing eighth in the World Championship, and while the Japanese rider Fumio Ito rode the BMW during 1960, this was the end for the RS. Not only was BMW restructuring, but the adherence to the Earles-type fork with its high steering inertia also limited the RS's competitiveness as a solo racer.

It remained a different situation in sidecar racing where the Earles fork was ideally suited. Walter Schneider and Hans Strauss won the Sidecar World Championship in 1958 and 1959. As the development of the racing sidecar proceeded towards lower outfits, only the BMW engine, driveline and suspension was retained. In 1960, Helmut Fath and Alfred Wohlgemuth won the championship on a private machine, while Scheidegger built his first kneeler outfit in an attempt to offset the power differential between his and the factory engines. The factory outfit of Max Deubel and Emil Hörner that gave them four consecutive world championships between 1961 and 1964 persisted with a conventional sitting position and the fuel tank above the engine. Scheidegger's machine on the other hand was a generation removed. By 1962, each wheel had a disc brake, and for 1964, he fitted 10in magnesium Mini car wheels at the front and side. When the factory withdrew from competition at the end of 1964, Scheidegger and John Robinson took their first World Championship, repeating this in 1966.

Even though the RS engine had long ceased production and spares dwindled, BMWs dominated the sidecar grids. After Scheidegger was killed at Mallory Park early in 1967, Klaus Enders (with Ralf Engelhardt and Wolfgang Kalauch) went on to win World Championships in 1967, '69, '70, '72, '73, and '74. With engines prepared by Dieter Busch, Enders final BMW kneeler outfit produced around 65bhp, featured a wide rear car tyre, and used a very short steering column with a U-link pivoted front fork. After 21 years and 19 World Championships, the RS monopoly ended, and the two-strokes took over. Although not exceptionally powerful, the RS BMW engine proved incredibly reliable and ideally suited to sidecar racing. Ironically, its era ended just as BMW was entering a new age of profitability, but von Falkenhausen wouldn't consider producing a two-stroke racer.

THE R50, R69 & R60

Production peaked in 1954 with 29,699 motorcycles manufactured, but already it was evident that the short-travel pre-war plunger frame was struggling to harness the moderate power of the R68. A new chassis was needed, and this arrived with the R50 and R69 of 1955. Unfortunately, the release of the R50 and R69 also coincided with a dramatic slump in motorcycle sales as cars became more affordable. Many of BMW's competitors (Horex, Adler, Ardie, and DKW) had vanished by the end of the decade, and BMW too was fortunate to survive. After dropping to 23,531 in 1955, and 15,500 in 1956, by 1957 production had crumbled to only 5,429, and most of these were for export markets.

The R50/R69 frame (245/1) was derived from the RS54 racer of 1953. This not only incorporated a swingarm rear suspension, but following racing experience the driveshaft housing connected to the swingarm. The front suspension was a development of the leading link swingarm type developed by Englishman Ernie Earles. By the early 1950s, swingarm rear suspension was considered de rigueur for a sporting motorcycle, but Earles-type forks were not as popular, except for sidecar use. Earles forks did

possess the advantage of not diving under braking, but they also put handling at a disadvantage because of the increased unsprung weight and steering inertia. Manufactured by BMW under licence from Earles, two tubes were angled back from the steering head to behind the front wheel. A swingarm pivoted on a fixed axle was attached to these tubes, with two hydraulic dampers connecting the swingarm to the lower fork crown. There were tapered roller bearings for both the front and rear swingarm pivots. Despite the addition of swingarm rear suspension the frame still resembled the earlier plunger type, but with supports for the swingarm and steel cups locating the tops of the shock absorbers. Like the RS54, the driveshaft was enclosed in the right side of the swingarm, with the universal joint moved to the gearbox end of the driveshaft. The wheels were also now 18in, and the toolbox was hidden in the left side of the fuel tank.

The engine of the R50 (M252/2) was essentially that of the R51/3, but with four-ring (rather than five-ring) pistons providing a slightly higher compression ratio (6.8:1), and two Bing 1/24/45 and 46 carburettors. The previous I-section con-rods now had sword-shaped shanks. The power was up slightly to 26bhp while the R69 (268/2) retained the higher-performance R68 engine, along with the barrel-shaped rear

The R50 incorporated the Earles
fork and rear swingarm of the
RS54, but the engine was very
similar to the R51/3's.
BMW Mobile Tradition

Opposite: One of the classic BMW
twins of the post-war era was the
sporting R69S. *Ian Falloon*

crankshaft roller bearing. New for both was the
gearbox and clutch. There was no external hand lever
and the four-speed gearbox featured an improved input
shaft shock absorber. There were also two sets of
gearbox ratios available, one for solo use and another
for sidecars.

With two Boge hydraulic dampers, the Earles-fork
R50 and R69 offered improved ride over earlier
models, and for 1956, these forks were adopted over
the entire range. The R60 replaced the R67/3 as a
sidecar hauler, sharing the same 28bhp engine

(M267/4), but with slightly different 24mm Bing
carburettors. Because of the deepening sales crisis,
there was virtually no development between 1955 and
1960. A bigger taillight appeared in 1957, but that was
the extent of progress.

The main reason for the stagnation of motorcycle
development was due to the spiralling losses caused
through the manufacture of the expensive eight-
cylinder 502, 503 and 507 cars. Although the small
Isetta was selling in reasonable numbers, profits on
these weren't enough to offset the loss of around

DM5,000 on each large car. In an effort to improve profitability, the Austrian distributor Wolfgang Denzel produced the prototype of a 700cc car based on the Isetta chassis, during 1958. Unlike the Isetta, this was a conventional car, with a 697cc fan-cooled version of the boxer motorcycle engine. Although it was received positively, with 30,000 advance orders, by the end of 1959 BMW faced bankruptcy. At a shareholder meeting on 9 December 1959 Denzel forestalled a management decision to sell BMW to Daimler-Benz.

Denzel found that the development costs of the 700 had been written off illegally and didn't appear in the balance sheet. Two shareholders, Eric Nold and Friedrich Mathern, managed to obtain 10 per cent of represented share capital to adjourn the meeting, and the deadline for the sale to Daimler Benz elapsed.

With the fate of the company now in the hands of the shareholders, attention turned to the major shareholder Herbert Quandt. The Quandts were one of Germany's richest industrial families, but Herbert's

There was little development of the
R50 and R60 after 1955, but a
slightly revised R60/2 was released
for 1960. *Ian Falloon*

THE R50/2, R60/2, R50S & R69S

Improvements to the R50/2 and R60/2 were primarily
aimed at increasing reliability. While they looked
similar to their predecessors, inside the engine (M252/2
and M267/5) had stronger bearing housings, clutch,
crankshaft, and improved crankcase ventilation. The
new three-ring pistons provided a 7.5:1 compression
ratio and a slight increase in power for the R60/2 to
30bhp at 5,800rpm. While many examples were fitted
with an old-fashioned solo pan-type saddle, a dual seat
was also available, and Hella bar-end turn signals were
fitted as standard.

Replacing the R69 was a more sporting R69S,
joined also by the R50S. Developments to the engine
(M252/3 and M268/3) included even higher
compression pistons (9.5:1), rockers pivoting on needle
bearings under two-rib valve covers, larger inlet ports,
Bing 26mm carburettors, and larger volume air filters
and mufflers. The four-speed gearbox now featured
closer ratios, and there was a hydraulic steering
damper to quell instability at the rather optimistic
claimed top speed of 175km/h (109mph), for the R69S.
With 42bhp at 7,000rpm, the R69S was the strongest
performing production BMW twin up to that time, but
with this performance came reduced reliability.

It did not take BMW long to respond to reliability
problems and by 1962 there were new cylinders and
pistons, and from September 1963, a rubber-mounted
vibration damper was incorporated on the front of the
crankshaft. While the anti-vibration damper now
meant the R69S could run all day at high rpm, it was a
source of irritation for owners as it required frequent
maintenance. The high revving R50S (35bhp at
7,650rpm) placed even more demands on the
crankshaft, despite the sports models featuring a
stronger rear spherical roller main bearing, and by
1962, it was dropped from the range. The chassis of
both S models (245/2) remained robust and heavy, and
at 202kg (445lb) the R69S was hardly a true sporting
motorcycle. The Earles forks also contributed to slow
and ponderous steering, and from early 1965, stiffer
springs were specified for solo use to improve the
handling.

During the late 1950s and 1960s there were
occasional racing successes for the Earles-fork twins,

interest in keeping BMW alive was more than
financial. Impressed by the commitment shown by the
smaller shareholders, workers, and dealers, he decided
to personally supervise the restructure of BMW.
Entrusting his personal legal advisor, Wilcke, with the
responsibility, the aircraft division was sold off and
there were now limited resources for development of
new products. Although motorcycles were lower on
the priority than cars, the existing motorcycle range
was improved and expanded.

The R69S lasted from 1960 through to 1969, this being the 1962 version. *Ian Falloon*

primarily in endurance events where their reliability gave them an advantage. London BMW dealer MLG entered an R69 in the 1958 Thruxton 500-mile (800km) race, and a fourth place encouraged further competition. In the hands of Peter Darvill and John Lewis, their R69 won the following year. Darvill and Bruce Daniels then pulled off a more amazing victory by winning the Barcelona 24-hour race at the twisting Montjuich circuit. They narrowly failed to win at Montjuich in 1960, but with factory assistance, the MLG R69 won again in 1961. Even by 1964, the BMW twin was competitive, with Darvill and Norman Price coming second this year. The success of the MLG racer led to them setting a 24-hour record at Montlhéry, near Paris, during 1961. A team of four riders, Sid Mizen, Ellis Boyce, George Catlin, and John Holder, covered 4,220km (2,622 miles) at a speed of nearly 176km/h (109mph), the record standing for 16 years.

Compared to British motorcycles of the 1960s, the BMW R69S appeared to be an anachronism. Instead of a tiny fuel tank and slim seat, the R69S offered a 17- or 24-litre tank, small solo saddle, or wide dual seat. Only available in black or other colours to special order they remained expensive and exclusive. Ultimately, the conservatism of the R69S counted against it during the 1960s. While the Earles forks undoubtedly assisted in its suitability for sidecar attachment, sidecars went out of fashion during the 1960s. As the automotive side of the company expanded motorcycle sales stagnated, and BMW's response was to release a series of three motorcycles exclusively for the US market, all with telescopic forks.

THE R50US, R60US & R69US

The impetus for fitting telescopic forks came from BMW's experimentation with them on the ISDT bikes following their tentative re-entry into off-road competition in 1958. A factory 246GS was produced during 1963, based on the R69S but with a higher mounted engine and telescopic forks. This led to Manfred Sensburg and Sebastian Nachtmann riding similar ISDT machines during 1965 and 1966. They were outclassed by the lighter two-strokes, but BMW learnt how to successfully produce long travel

telescopic forks. For the US models of 1967, these forks were simply grafted on to the existing R50/2, R60/2 and R69S, although the sidecar lugs on the frames (245/3) were removed, as the telescopic forks were considered unsuitable for such an attachment. The engines were unchanged for the US versions, and all came with a 4.0 x 18in rear tyre.

For their day, the leading axle telescopic forks were very sophisticated, with progressive rebound and compression damping achieved through a tapered hydraulic metering rod. The only change from the ISDT fork was stiffer springing, and they provided a considerable, 214mm of travel. Compared to the Earles fork the telescopics were lighter, offered improved high-speed performance on bumpy roads, but heavier low-speed steering. They also increased the wheelbase slightly. The telescopic forks certainly presented a more modern image, but it was only a stopgap. Production levels gradually decreased and by 1969, only 4,701 motorcycles were manufactured.

By the end of the 1960s, the R50/60/69S didn't appeal to the average motorcyclist who demanded more power, noise and flashiness, even if it did mean putting up with vibration and a lot of maintenance. BMW's stood apart from the mainstream. Not only did the market require a new machine, the 18-year-old engine, with its built-up crankshaft, expensive ball and roller bearings, and gear-driven camshaft, was uneconomic to produce in quantity. To make more power would also require considerable modification to maintain reliability, so an easier solution was to create an all-new engine. The R69 engine's demise on 13 May 1969 coincided with the end of motorcycle production at Munich.

This 1979 R100S continued the tradition of the classic R90S, but featured cast-alloy wheels and a rear disc brake. *Ian Falloon*

5 A NEW GENERATION OF BOXERS

During the 1960s, car production increased, while motorcycle sales stagnated and BMW could have elected to cease motorcycle manufacture and concentrate on ventures that were more profitable. There had been little development on their motorcycle range for a decade although the motorcycle market had changed dramatically during this period. Not only had the European motorcycle market collapsed, but the Japanese manufacturers had begun producing powerful, reliable, and sophisticated, larger displacement machines. Against this background of pessimism, the technical director, Helmut Werner Bönsch, then managed to launch a new series of motorcycles.

Bönsch knew that for BMW to remain a viable motorcycle producer they had to expand their market for quality, luxury touring motorcycles, and saw the future as being with the development of the traditional flat-twin. Aimed at the rider who placed a premium on comfort and convenience, the new /5 series was the most radical motorcycle design in BMW's history, and would grow to become one of the most successful.

All that was needed for the production of the /5 was a suitable factory, and BMW decided to convert their repair and machine work facility in Spandau, West Berlin. Siemens & Halske had set up an aircraft engine factory at Spandau in 1928, and in 1936, it became Brandenburgische Motorenwerke (Bramo). In 1939, Bramo merged with BMW to become BMW Flugmotoren-Gesellschaft. After the war, the works became BMW Maschinenfabrik Spandau, and from 1958 it was involved in the manufacture of vehicle components. Motorcycle production commenced at Spandau in September 1969 with the R60/5. A month later, the R75/5 joined it, and in November, the R50/5 completed the line-up. By the end of the year, 1,205 motorcycles had left the Spandau works.

THE R50/5, R60/5 & R75/5

Von der Marwitz may have adhered to the traditional 'boxer' layout, but the engine (designation M04*) was all new. Three displacements were offered, 498cc, 599cc, and 745cc, all sharing the same 70.6mm stroke and basic architecture. There were a number of

HANS-GÜNTHER VON DER MARWITZ

Continuing the tradition of Rudolf Schleicher and Alexander von Falkenhausen, Hans-Günther von der Marwitz was an enthusiastic motorcyclist. Hailing from a distinguished family, von Marwitz was born in 1927 and joined BMW in 1964 from Porsche. He was used to racing around on an AJS 7R, and was dismayed at the handling of the Earles-fork BMW. When he was entrusted with the design of the new BMW motorcycle, von der Marwitz wanted a machine to handle as well as a Manx Norton.

significant departures from the earlier engine. Inside the one-piece aluminium, internally reinforced tunnel housing crankcase was a one-piece forged crankshaft (without a centre bearing to minimise cylinder offset) running in plain bearings. As was envisaged back in 1932 with the M205 to M208 designs, the camshaft was now underneath the engine. A duplex chain drove this from the front of the crankshaft. Many components came straight off the automotive production line; the automatic tensioning camchain was from the rally-winning sports saloons, and the three-layer plain bearings for the crank and con-rods from the six-cylinder cars.

As the bearings required high-pressure lubrication, an Eaton trochoidal oil pump was fitted at the rear of the camshaft. At the front of the crankshaft was a three-phase 180-watt alternator to power the new 12-volt electrical system. The earlier magneto went in favour of battery and coil ignition with an automatic advance, and above the engine (on the R60 and R75 – optional on the R50) was an electric starter motor. As the flywheel and single-plate dry clutch were an automotive type the electric start was a simple and effective addition.

Although the valve actuation system retained pushrods, with the pushrod tubes now underneath the cylinders the engine looked more modern. Aluminium was used extensively to minimise the effect of the heavy starting system, including alloy instead of steel cylinder barrels, with a cast-iron sleeve bonded to the cylinder through the Al-Fin process. This contributed to increased cooling efficiency.

The cylinder heads were also new, with a much shallower included valve angle of 65°. The R75/5 had

large, 42mm and 38mm valves, and a more radical camshaft than the R50/5 and R60/5. The R75/5 also had 32mm Bing constant velocity carburettors as opposed to the Bing 26mm concentric carburettors on the smaller versions. All this contributed to 50bhp at 6,200rpm for the R75/5, although the R50/5 and R60/5 made a very modest 32bhp and 40bhp respectively.

The four-speed, three-shaft gearbox bolted on the rear of the engine, and apart from strengthening, the drivetrain was similar to the previous model. What was new was the backbone-type, double-loop frame designed exclusively for solo riding. There were no sidecar lugs, and the frame was constructed of variable section tapered and oval tubing with a bolted-on rear subframe. BMW also claimed to have incorporated some controlled longitudinal flexing at the steering head to improve handling. The weight of the frame was reduced from 17.5kg (38.6lb) to 13kg (28.7lb) and further weight saving extended to the wheels and brakes. There were light alloy wheel rims (a 19in on the front), and narrower (200 x 30mm) drum brakes with special linings which had been developed for the last Porsche cars to have drum brakes.

Suspension included the telescopic forks of the earlier US versions, providing a generous, 208mm of travel, with twin shock absorbers at the rear giving 125mm of travel. Practical features extended to a large, 24-litre fuel tank and a good-sized dual seat. With a host of lightweight features extending to glass-fibre mudguards, the new /5 series were also reasonably light for their class, the R75/5 topping out at 210kg (463lb).

The /5 certainly vindicated Bönsch's optimism. The R75/5 may not have been as powerful as the new Honda 750 Four, but it was no longer a staid and stodgy motorcycle only for the initiated diehard. For a rider interested in long distance, comfortable high-speed travel there was simply no other contender in 1969. Here was a motorcycle that could reliably cruise all day at 160km/h (100mph), with all the conveniences expected of modern machinery. Offering excellent handling, and respectable performance, the new boxer, especially the R75/5, was immediately successful.

As it was an all-new model there were many developments to the /5 series over the next three years. For 1971, there was a lower final drive ratio (10:32 instead of 11:32) to aid fourth gear performance, and

the carburation was altered to quell low-speed running problems. More careful assembly of the long-travel front fork, with closer tolerances, also alleviated some of the criticism of head shaking and wobbles.

Styling changes were the most evident development for 1972. For this year, there was the controversial 17-litre 'toaster' tank with chrome panels, and chrome-plated side panels to hide the battery. Primarily for the US market and so-called because of its apparent similarity to a kitchen appliance, this styling was attributed to then BMW sales director Bob Lutz, an American, but also a motorcycle enthusiast. It proved unpopular and for 1973, the smaller tank was standard, without the chrome panels. The 24-litre tank

was now an option. More colour schemes were available as the years went by, culminating in six colours by 1973. The days of black-only BMWs were now long gone.

Also new for 1972 was a wider rear wheel rim (WM3), longer and stronger front fork springs, stronger crankshaft and bearings, and new rocker shaft supports to reduce noise. The most important development occurred in January 1973 when the swingarm was lengthened by 50mm to improve the handling. This also allowed for a longer seat and room for a larger battery. Although the /5 series, and in particular the R75/5, re-established BMW as one of the world's leading motorcycle manufacturers, by 1973

The R75/5 was the first official post-war BMW 750. This 1973 version had a longer swingarm than previous examples.
Ian Falloon

the /5 needed updating. Certain features, such as the four-speed gearbox, drum front brake, and instruments in the headlight, harked back to an earlier era. On 28 July 1973, only three days after the 500,000th BMW motorcycle (an R75/5) came off the production line, the last /5 left the Spandau factory. The time was ripe for a BMW Superbike.

THE R90S

When the new /6 series was unveiled in September 1973, all the expected improvements were incorporated, but the new top-of-the range sporting R90S took the world by storm. Out of the R75/5 BMW created a cutting-edge sporting motorcycle, and the R90S proved unsurpassed on the road

performance, as well as setting new aesthetic standards.

The engine was shared with the rest of the /6 series, incorporating a five-speed gearbox with a revised gearshift mechanism, and stronger crankcases. This required reducing the size of the front counterweight to pass through the strengthened tunnel so tungsten plugs were inserted in the crank webs to restore the balance. The R90S featured higher compression, 90mm pistons (9.5:1), and feeding the engine was a pair of the new generation of 38mm Dell'Orto slide carburettors with accelerator pumps. There were black cylinders for improved heat dissipation, but otherwise the engine was identical to the R90/6, itself very similar to the R75/6 except for the larger pistons, and 40mm exhaust valves (up from 38mm). These few modifications had a significant effect on the power output, and the R90S

produced 67bhp at 7,000rpm. In the space of only four years, the power of the top-sporting boxer had increased by 63 per cent. The R90S was also easier to live with, a 280-watt alternator and larger battery providing more reliable starting.

It wasn't sheer power that set the R90S apart. The Kawasaki Z1 was more powerful, but the R90S was the first production motorcycle to feature a factory-fitted fairing as standard equipment. Designed by Hans Muth, the small cockpit fairing incorporated full instrumentation, including a clock and voltmeter, and complemented a beautifully styled fuel tank and seat. Each bike was hand-painted in Smoke Black so no two were identical. The R90S was also the first BMW to feature dual stainless-steel 260mm front disc brakes, although the floating single-piston ATE brake calipers

lacked ultimate power. To tidy the cockpit the front brake master cylinder was also located under the tank, as was a three-way adjustable hydraulic steering damper. While the frame was essentially identical to that of the long wheelbase R75/5, there was additional gusseting around the steering head. With its shaft drive and long-travel suspension, the R90S may have lacked the sharpness and handling precision of a comparable Ducati or Laverda, but more than made up for it in civility. It was on the road that the R90S excelled, and for its day it was a genuinely fast motorcycle, capable of 200km/h (125mph).

So good was the R90S there were very few developments for 1975. New switches, a faster action twist grip, and drilled discs complemented the choice of either Smoke Black or Daytona Orange. The

Helmut Dähne in action on a
semi-works R75/5-based Formula
750 racer in the 1973 Imola 200.
Mick Woollett

RACING /5s

Although the R69S had isolated success in long-distance racing during the 1960s, the R75/5 was a much more suitable basis for a racer. The R69S was difficult to set up for racing and required a specific riding approach to attain its best. Von der Marwitz may not have succeeded totally in creating a motorcycle that handled as well as a Manx Norton but it was surprisingly close, and by 1971, Helmut Dähne was achieving some good results in production racing in Germany. Hans-Otto Butenuth rode a special racer in the 1971 Production TT, finishing a creditable fourth, Dähne repeating this in 1972 and 1973. With the advent of Formula 750 in 1972, Butenuth, Dave Potter, and Dähne rode F750 machines in the Imola 200, Dähne finishing 13th. He campaigned the F750 bike during 1973, finishing 14th in the Imola 200.

On the other side of the Atlantic, Butler & Smith, the US distributor, decided to allow their service manager Helmut Kern to prepare a racing R75/5. Sales of the R75/5 were rather slow in the USA, and with some backdoor help from contacts at the

factory, Kern was able to obtain many special engine parts, such as titanium valve covers. The modified R75/5 engine was placed in a special German GP-style frame prepared by Volker Beer and Kurt Liebmann raced it on the East Coast and in Canada during 1972. Then, at the instigation of ex-patriot British rider Reg Pridmore on the West Coast, Kern approached Rob North to build a frame similar to those of the factory Triumph three-cylinder racers.

North built two frames, and the 745cc engine featured bigger valves, and 36mm racing Mikuni carburettors. With Ceriani forks and later, Lockheed disc brakes, Pridmore and Liebmann campaigned the Butler & Smith racers throughout 1973 and 1974. Weighing around 140kg (309lb) and capable of 265km/h (165mph), the handling was impeccable, but as they were running against two-stroke 750s in AMA racing, there were no race victories. This would change with the advent of Superbike racing and the R90S.

Below: Kurt Liebmann rode this /5-based racer in the 1972 Daytona 100-mile Junior race, but didn't finish. *Mick Woollett*

Below: At Daytona in 1973, Reg Pridmore (84) joined Liebmann (57) on the Butler & Smith racers. *Mick Woollett*

Bottom: Pridmore in action on the outclassed Rob North-framed racer in the 1974 Daytona 200. *Mick Woollett*

Opposite: Few motorcycles have
garnered the following of the
R90S, and it is well deserved.
This is the author's 1976 example.
Ian Falloon

Following on from the R75/5
was the R75/6, with a front disc
brake and five-speed gearbox.
Ian Falloon

pinstriping was now hand-painted instead of the tape
as on earlier models. Early gearbox shifting problems
were rectified through revised shifting forks, and there
was a more powerful starter motor. More modifications
were incorporated for 1976, most as a prelude for the
/7 series released later that year. To improve intake
flow and exhaust heat dissipation, the included valve
angle was reduced slightly. There were aluminium
pushrods and shorter and re-angled rockers. A stiffer
camshaft spindle resulted as the diameter was increased
from 12mm to 20mm, and the crankcases and front
main bearing were upgraded. A deeper sump also kept
the oil further away from the crankshaft, reducing oil
drag. The front fork travel was reduced slightly (to
200mm), with a firmer action. The swingarm was now
pressed and welded around its entire section for

additional strength and the front brake calipers and
master cylinder received larger pistons.

More than any other model, the R90S created a
performance aura for BMW that previously didn't
exist, even with the R75/5. With nearly 17,500
manufactured, they weren't exactly a rare limited
edition, but the R90S was BMW's first Superbike, and
one of the most beautifully styled and balanced of all
the boxer twins. Many dreamed of owning an R90S in
the early 1970s, but it was one of the most expensive
motorcycles available and unaffordable for many. Now
it is a much sought-after model, and rightly so, as an
R90S still provides a wonderful riding experience.
Later twins may be functionally superior, but the R90S
has rightfully earned its status as the classic BMW
motorcycle of the modern era.

SUPERBIKE AND PRODUCTION RACING

Many of the performance developments of the R90S emanated from the specially developed R75/5 campaigned by Helmut Dähne. Dähne and the BMW were particularly suited to the Isle of Man. In 1974, he came third in the Production TT and was leading the race in 1975 until a stone punched a hole in the well-worn rocker cover. Dähne finally gave BMW a victory in 1976, when sharing a factory-assisted machine with Butenuth, he won the ten-lap 1,000cc Production TT.

As production-based racing grew in the USA during 1974, Reg Pridmore rode one of the new R90Ss. After narrowly failing to win at Laguna Seca, when an ignition coil wire broke, Pridmore gave BMW their first production race-win, at Ontario. He was so far in front of Yvon DuHamel's second-placed Yoshimura Kawasaki that DuHamel assumed he had dropped out. Although Pridmore also campaigned the R90S during 1975, finishing 31st in the Daytona 200, the creation of the Superbike class for 1976 opened a new door for the R90S.

Superbikes allowed for much more engine modification, but retained standard frames. As the Japanese bikes were powerful, but ill handling, the regulations played into the hands of the European manufacturers, particularly BMW. Butler & Smith had the bikes (R90S), the budget, and the experience to produce an unbeatable Superbike. They may have looked like stock R90Ss, but the 1976 Butler & Smith Superbikes were extraordinary racers.

In Norwood, New Jersey, Udo Gietl, assisted by Todd Schuster and Tom Woods, looked closely at the regulations and stretched the boundaries to the limit. The engine was modified to not only increase horsepower, but also improve ground clearance. Borrowing from the factory ISDT engines, shorter rods and barrels were possible through short, 95mm Venolia pistons with the gudgeon pin inside the ring grooves, and with 10mm shorter Carrillo con-rods. This enabled the engine to be 28mm shorter on each side. With 995cc, the compression ratio was 12.2:1. Kenny Augustine worked the cylinder head with 46mm and 39mm valves, and twin spark plugs, while Sig Erson ground a special camshaft. With the Dell'Orto carburettors bored out to 40mm, the eventual power was around 100bhp at 8,300rpm.

The engine may have used standard American hot-rod technology, but Gietl really stretched the regulations by extensively strengthening the swingarm and changing the twin shock absorber set-up to incorporate a single Koni Formula One car damper. Further chassis modifications involved moving the engine and driveshaft to the right to allow for a racing slick tyre, and moving the engine forward by 30mm and up by 10mm for improved weight distribution and ground clearance. Weighing 204kg (450lb), the top speed was in the region of 235km/h (146mph).

At the opening Superbike race at Daytona the appearance of the three Daytona Orange Butler & Smith BMWs left the opposition stunned. Nicknamed the 'Bavarian Murder Weapons' by Schuster, in the hands of three of the best Superbike riders in America, Reg Pridmore, Steve McLaughlin and Gary Fisher, the

Left: Steve McLaughlin, side-by-side with Reg Pridmore, on the special Butler & Smith R90S monoshock Superbike. McLaughlin, here on his way to victory in the 1976 Daytona Superbike race, headed a BMW one-two. *Mick Woollett*

Below: This is McLaughlin's 1976 Daytona-winning R90S Superbike, one of the most successful of all BMW racers. *Mick Woollett*

Opposite: Reg Pridmore rode this stock-looking R90S in the 1975 Daytona 200-mile race. *Mick Woollett*

BMWs ran away with the Superbike race. In a photo finish, McLaughlin narrowly beat Pridmore, who had reverted to a twin-shock set-up for the race. With victories at Laguna Seca and Riverside, Pridmore went on to give BMW their first and only AMA Superbike Championship.

Butler & Smith withdrew from racing after Daytona 1977, and Gietl and Schuster took up the racing programme independently. Superbike regulations for 1977 banned the monoshock, and although there were now dual-piston Lockheed brakes and engine developments increased the power by around 6bhp, the BMWs were outclassed. More development for 1978 though, saw the BMW almost take out the AMA Superbike Championship. Gietl worked his magic on the still R90S-based racer and in the hands of John Long it finished third at Daytona. Consistent results throughout the season, and a win at Mosport, ended with Long finishing a close second in the championship. Although this represented the end of competitive racing in Superbike, the boxer twin continued to hold its own in the Battle of the Twins series, with Stuart Beatson winning in 1983.

Opposite: Following on from
the R90S was the equally
ground-breaking R100RS of
1976. *Ian Falloon*

THE R60/6, R75/6 & R90/6

Alongside the R90S was a revamped /6 series for 1974. The underpowered and unpopular 500 was dropped altogether, and the range expanded upwards to include the R90/6. Compared with the R90S, the R90/6 engine featured lower compression (9:1) 90mm pistons, and Bing 32mm constant-velocity carburettors. The power was down to 60bhp at 6,500rpm. Almost identical to the earlier R75/5 and R60/5, but for the five-speed gearbox, were the smaller R75/6 and R60/6. They all shared similar styling to the final /5 series, but there were individual instruments and the R90/6 and R75/6 received a single front disc brake. The /6 series models continued the BMW tradition of offering superb touring ability and useable, if moderate, power. As the /6 series soon earned a reputation for outstanding build quality and reliability, sales were strong. For 1975 the larger models received a drilled disc, there was new switchgear, and for 1976, they shared the developments of the R90S. Although overshadowed by the R90S, the /6 series represented superior value for money, and sold in greater numbers. The /6 series were exceptional motorcycles, and many examples have covered hundreds of thousands of miles with few problems.

In 1976, coinciding with the new engine developments, there was a reorganisation of BMW's motorcycle division into a company separate from the cars. The introduction of the successful /6 series resulted in production accelerating from 15,078 in 1973, to 23,160 in 1974, and 25,566 in 1975. Despite losses in the USA, exports were increasing, particularly in Britain, Italy, and the Netherlands. After fighting for survival only a decade earlier, the BMW motorcycle had staged a miraculous comeback. However, lulled into complacency, this success was short-lived and a sales slump during 1978 and 1979 saw almost the entire management team replaced. The new team realised the need for innovation and rose to the challenge.

THE R100RS

If the R90S signalled that BMW was prepared to expand beyond their traditional horizons, the R100RS added a further dimension. Creating a vague association with the racing RS54, the R100RS didn't feature a new double overhead camshaft engine, but was the first series production motorcycle to be fitted with a full touring fairing. Wind tunnel-designed by Hans A. Muth, the nine-piece fairing featured an integrated cockpit and spoiler to reduce front wheel lift by 17.4 per cent. Air resistance was reduced by 5.4 per cent but the effectiveness of the fairing was particularly impressive. With low sporting-style handlebars, the R100RS was the most efficient high-speed touring motorcycle available.

Complementing the fairing was a 980cc engine (M65*), with new, 94mm pistons. Instead of the Dell'Orto 38mm carburettors of the R90S, there was a pair of Bing V94 constant-vacuum 40mm carburettors. Inlet valve diameter went up to 44mm, exhaust pipes up to 40mm, and the power to 70bhp. Because of the full fairing, most engine developments, many introduced for 1976, were aimed at reducing noise. There were now shorter and thicker cooling fins, O-rings to seal the cylinder bases, and a more effective crankcase ventilation system. The old-style rocker covers made way for new angular, heavier, black anodised ones, and there was a stronger clutch and revised starter ratio. The kickstart was no longer considered essential, but remained an optional extra. A smaller and lighter 240-watt alternator was also fitted to the R100RS (and R100S) to provide clearance at high rpm with more crankshaft whip, between the magnets and stator.

It wasn't only the engine that received tweaks for the R100RS. The chassis, with the stronger 1976 swingarm, featured an additional brace welded between the two front frame downtubes. The suspension was revised to provide a more progressive ride, and the RS received blue anodised front brake calipers. There was a choice of a single or dual seat, and wire-spoked or heavier cast-aluminium wheels. The rear 200mm drum brake on the cast wheel incorporated air scoops.

Although one of the most expensive motorcycles available, the R100RS immediately garnered a strong following. This wasn't surprising considering its capability. Not only did the fairing work superbly, but the R100RS was also fast (around 200km/h), reliable, good handling, and its 24-litre fuel tank provided a near 400km range. No other motorcycle could match it as a long-distance sport-tourer. For 1978, the wheels were cast aluminium only, with a rear disc brake and

This 1979 R100S continued
the tradition of the classic R90S,
but featured cast-alloy wheels
and a rear disc brake. *Ian Falloon*

Brembo caliper, and there was a revised gearshift and
linkage. The front brake calipers were no longer
anodised blue.

The colour range expanded during 1978 from the
original silver, with a fade-prone gold also available.
There was also the first of multiple limited editions, the
white Motorsport. These were followed by a handsome
silver and Royal Blue for 1979. Engine developments
this year included a single-row camchain with a spring-
loaded and hydraulically damped tensioner, while the
driveshaft was fitted with a ramped coupler-type shock
absorber. Of all the developments aimed at improving
the gearshift over the years, this was by far the most
successful. While the engine specifications for most
markets continued unchanged for 1980, all that year's
US R100s included an air suction emission system and
a lower 8.2:1 compression ratio.

There were more changes for 1981 (shared with all
R100s), the engine series now designated the A10 (with
an internal designation of R246/247). Nikasil cylinders
replaced cast-iron liners, saving weight and increasing
longevity, and there was a new diaphragm clutch
spring, a 4.5kg (10lb) lighter flywheel (sheet steel
instead of cast), and a 40 per cent easier clutch action.
The sump was shorter and deeper, increasing capacity,
and there was a new air filter, the rectangular flat
paper element now in a black plastic box with two
removable plastic snorkels. Finally, there was
breakerless electronic ignition, and the more-powerful,
280-watt alternator returned.

To reduce noise levels while maintaining 70bhp, the
exhaust system incorporated two balance pipes, while
US versions received exhaust air-port injection to clean
up emissions. Also new were Brembo front brake

The short-lived R75/7 was
replaced by the R80/7 during
1977; all /7s featuring the
attractive R90S-style fuel tank.
Ian Falloon

calipers, with the Magura master cylinder moved from underneath the fuel tank to the handlebar, accompanied by a new fork that incorporated some Teflon components, although fork travel remained the same. There was a stronger die-cast rear axle housing (intended for the forthcoming Monolever R80 G/S), and a stronger swingarm. The colours were Smoke Black or red, with an optional oil cooler (standard in some markets). With the standard oil cooler the centre section of the fairing was closed off. However, this increased heat build up and caused electrical problems. The year marked the end of further developments to the R100 series until its demise in 1984. All resources were now involved in the development of the new K series and although a 'Last Edition' white R100RS and blue and silver Series 500 became available in 1984, popular demand would see its return only two years later.

THE R100S & R100CS

Although the R100RS assumed the status of range leader in 1977, the earlier R90S style continued with the R100S. For 1977, the 70bhp, 980cc engine was reserved for the R100RS, while the R100S had a slightly lower rated (65bhp) engine. Still with Bing 40mm carburettors, the R100S had smaller diameter exhaust pipes (38mm). All other engine and chassis improvements were shared with the R100RS, and despite more weight and less power than the earlier R90S, the R100S provided outstanding performance. This was ably demonstrated in the 1977 Australian Six-Hour production race. R90Ss finished third in 1975, and second in 1976, but for 1977 the local distributor hired four top riders to meet the challenge of the Kawasakis and Ducatis. Ken Blake teamed with veteran Joe Eastmure, while Helmut Dähne was flown in to pair with Tony Hatton. The pair of new R100Ss dominated, with Blake and Eastmure winning while Hatton and Dähne came third. Considering the race was for unmodified production motorcycles it was a spectacular performance.

Cast wheels and a rear disc brake made it to the R100S for 1978, with the R100RS 70bhp engine introduced for 1979. Then, for 1980, with sales stagnating compared with the R100RS, BMW introduced the almost-retro R100CS. Retaining the 70bhp engine, this was first displayed with wire-spoked wheels, but most production examples had cast wheels, with the Simplex single leading-shoe rear drum brake. Several versions were offered to stimulate sales, including a silver 'Outback' edition for Australia (with wire-spoked wheels). Despite its retro appeal, the R100CS was not particularly popular, and only around 4,000 were produced between 1980 and 1984. It is now considered one of the more collectable modern boxer twins.

THE R60/7, R75/7, R80/7, R100/7, R100/T & R100

There were three new /7 series models for 1977, and as with the preceding /6 series they were all visually similar. As before, the R60/6 featured smaller valves, Bing slide carburettors, and a milder camshaft, but now sported a single front disc brake instead of the previous drum. With the 24-litre R90S fuel tank and angular valve rocker covers, the styling was more modern, but it wasn't enough to disguise the fact that the R60/7 was still a 40hp, 215kg (474lb) motorcycle. Performance was barely adequate, and the under-stressed R60/7 was really a machine more suitable for diehard enthusiasts interested in long life rather than performance. As it was produced primarily for police forces requiring a budget-price BMW, the R60/7 was dropped for 1979, effectively replaced in this role by the R80/7 and later, the R80 Monolever.

While the R75/7 shared similar engine specifications with the R75/6, the R90/6 grew to become the R100/7. With a slightly lower compression ratio (9:1), smaller intake valves (42mm), and Bing 32mm carburettors,

the R100/7 was 5hp shy of the R100S. The R75/7 still offered possibly the best balance between power, mid-range torque and smoothness, but more buyers preferred the 980cc R100/7. In an effort to widen the range appeal the R75/5 made way for the slightly larger R80/7 during 1977.

More displacement for the R80/7 (M85*) came through 84.8mm pistons, also bumping the compression up slightly to 9.2:1, and aiding in a power increase to 55bhp at 7,000rpm. A lower compression (8:1) 50bhp version was also available for markets that could only obtain low octane fuel. Shortly afterwards, for 1979, the R100T replaced the R100/7. In an effort to rationalise specification, the R100T featured the 65bhp engine with higher compression and 40mm carburettors, and cast aluminium wheels (but with a rear drum brake instead of the disc of the RS, S and RT). After only two years, the R100 replaced the R100T. Incorporating all the improvements of the 1981 Model Year, the R100 featured a lower compression (8.2:1) 980cc engine, but retained the 40mm carburettors. The power was rated at 67bhp at 7,000rpm.

THE R100RT & R80RT

Following the success of the R100RS, BMW decided to expand the luxury motorcycle concept to include a full-dress touring machine. Introduced for 1979, the R100RT was aimed squarely at the lucrative US market. It was not only the most lavishly equipped BMW ever, it was also the most expensive motorcycle available in America. Sharing its 70bhp 980cc engine and chassis with the R100RS, the R100RT added a full fairing with a tall screen to complement the high and wide handlebars. The screen was cleverly provided with adjustment for four different heights, and four rake angles. Lockable compartments were incorporated forward of the fuel tank, and there were air vents on either side. With Krauser saddlebags as standard equipment the touring specification was impressive, but the extra weight taxed the engine and suspension. No longer an exceptional handler, and certainly no dragstrip tearaway, the R100RT's sales were disappointing, especially in the USA. When the compression ratio was lowered on US versions for 1980, the performance deteriorated further.

BMW endeavoured to make amends for 1981. New

THE R45, R65 & R65LS

The demand for a cheaper, more compact, smaller capacity motorcycle led to the release of the R45 (M76*) and R65 (M84*) in 1978. Replacing the R60/7, but closely modelled on the bigger twins, these were either intended for specific insurance categories (such as the 27hp class in Germany), or as an entry-level model that would hopefully lead to the purchase of a larger twin. Although the crankcase castings were the same as for the larger versions, the R45/65 had a shorter-stroke crankshaft (61.5mm) that allowed the engine to be 56mm narrower. Also different was a single-row roller timing chain with oil-damped tensioner, and the ignition points cam was no longer incorporated with the camshaft. There was a lighter flywheel, and smaller clutch to encourage revs. The five-speed gearbox was shared with the larger twins, but there was a lower final drive ratio. Ultimately, the smaller R45 and R65 lacked power, especially when compared with cheaper Japanese middleweights, and even the R65 was afflicted with mid-range vibration and stumbles.

The R45/65 frame was similar to the R80 and R100 models, but did not feature oval-section downtubes or additional gussets. With a 50mm shorter swingarm and new, shorter-travel centre-axle forks, the wheelbase was 65mm less. Also new was an 18in front wheel and special twin-piston brake caliper for the single front disc. A smaller angular-shaped fuel tank (22 litres) and lower seat height of 770mm contributed to a physically smaller machine, but the weight was only 10kg (22lb) less than an R100.

Over the next few years there were many minor developments to the smaller boxer series. After 1980, the engines were designated A20 (internal designation R248), and there was a lighter clutch, Nikasil cylinders, breakerless ignition and two connecting crossover pipes in the exhaust system. There was also a larger oil pan, additional cushioning in the driveshaft, and a 10mm longer swingarm. The R65 received larger valves (40mm and 36mm), and the power was increased to 50bhp (from 45bhp). To match the increased power there were optional dual front disc brakes, but these developments did little to endear the R45 and R65 to prospective buyers.

Below: As a styling exercise the R65LS was reasonably successful, but it wasn't enough to save the smaller boxer. *BMW*

Opposite: While not as popular as the larger twins, the R65, like this 1979 model, was an extremely competent and well-balanced motorcycle. *Cycle World*

In an effort to widen the appeal of the series, the sporting R65LS was offered for 1982. Sharing its 50bhp engine with the R65, the R65LS was primarily a styling exercise that incorporated a cockpit with an integrated spoiler. Claimed to reduce front wheel lift by a third, the accentuated styling continued through to the seat with its integrated grab handles. The exhaust pipes were black chrome, and while the R65LS included dual front disc brakes, the most worthwhile feature was the new wheels. Jointly developed with Alusuisse, the front rim width was increased to 2.15in, yet was lighter than before. Undoubtedly a competent sporting middleweight, the R65LS was viewed by most BMW traditionalists as a cosmetic adulteration and was a poor seller. Unloved, and probably under-appreciated, the small twins were discontinued during 1985. When the next R65 appeared, its basis was the new R80.

Boge Nivomat load-levelling rear shock absorbers and revised fork damping improved the ride and handling, while other developments were shared with the R100RS. In 1982, the R100RT was joined by the R80RT. In an effort to overcome stagnating R100RT sales, the R80RT was almost an identical motorcycle, but with the R80/7 engine, a rear drum brake, and without standard luggage or Nivomat rear suspension. Significantly, it sold for a lot less than the R100RT, even though the performance was decidedly leisurely. The 50bhp engine struggled to power the 235kg (518lb) R80RT beyond 100mph (161km/h). Despite this, the R80RT proved popular, and when the revised boxers were offered in 1984, the R80RT survived. It wasn't until 1987 that the R100RT would make a return.

MONOLEVER TWINS THE R80 G/S & R80ST

The successful re-entry into the world of off-road competition during 1979 prompted the development of the dual-purpose R80 G/S. Indicating Gelände/Straße (off-road/street), initially it seemed incongruous that an off-road motorcycle could displace 800cc, incorporate shaft-drive, and weigh 186kg (410lb). Based on the street R80 and R65, the R80 G/S was certainly too large and heavy to be a serious off-road motorcycle, but BMW's engineers managed to create one of the best handling and most effective street BMWs ever. The R80 G/S successfully established a niche market, and created another classic motorcycle that would garner a following to match the R90S and R100RS.

The idea for the R80 G/S came from Rüdiger Gutsche, a suspension engineer and ISDT veteran. Gutsche had already produced a special one-off /5 series enduro machine back in 1975, and in 1979 caused a stir as a marshal in the ISDT when he rode a further-improved version. This machine influenced the design of the R80 G/S, and other members of the development team built similar machines for participation in the traditional Dolomite Rally in the Italian Alps. Using the engine of the R80/7, the R80 G/S incorporated the 1981 series improvements such as Nikasil cylinders, electronic ignition, and lighter flywheel for easier clutch action. The 800cc engine was placed in an R65-derived frame with a Monolever swingarm and single shock absorber. Claimed to provide 50 per cent more torsional rigidity, along with a 2kg (4lb) weight saving, the Monolever soon found its way on to the rest of the BMW motorcycle range.

As it was intended as a dual-purpose motorcycle, the R80 G/S had a 21in front wheel, long-travel suspension (200mm from the front 35mm forks, and 170mm at

Below: One of the most popular variations on the boxer theme was the R80 G/S. This featured the new Monolever swingarm and was a surprisingly competent street bike. *Ian Falloon*

Opposite top: To celebrate the victories in the Paris–Dakar Rally, this Paris–Dakar version of the R80 G/S was produced in 1985. *Australian Motorcycle News*

Opposite bottom: Although the R80 G/S was a resounding success, the street R80ST was a commercial failure, and lasted only two years. *Cycle World*

the rear), and high plastic mudguards. Braking was by a single 260mm disc with Brembo caliper, and the ubiquitous Simplex drum. The power, weight and top speed of around 168km/h (104mph), required specific dual-purpose tyres developed by Metzeler. After a development period of 21 months, the R80 G/S debuted in the autumn of 1980.

There was no doubt the R80 G/S was a brilliant motorcycle, and unlike the R45 and R65, exceeded all initial expectations. With over 6,000 produced in 1981, 20 per cent of BMW's motorcycle sales were the R80 G/S that year. Production continued through until 1987, with an optional 32-litre Paris–Dakar fuel tank, red solo seat, and a luggage rack available from 1984. The tank carried a Gaston Rahier signature and included twin fuel petcocks.

So successful was the R80 G/S, particularly as a street bike, that BMW decided to capitalise on this in 1982 with the R80 ST. The R80 ST was essentially a parts bin special, and an amalgam of the R80 G/S, R65 and R100. The basic engine and chassis was that of the R80 G/S, but with a shorter shock absorber, wider rims

The GS80 of 1979 was the precursor to the Paris–Dakar machines.
Stefan Knittel

PARIS–DAKAR SUCCESS

Off-road competition, particularly ISDT, was important for BMW throughout the 1930s, and after the Second World War. However, during the 1960s, without a suitable machine as a basis, BMW only supported these events half-heartedly. That all changed in 1973 when special 750cc machines were prepared for the ISDT in the USA. Based on the R75/5, the machines were outclassed, but they led to a special GS80 for 1979. With 95mm pistons and a 61.5mm R65 crankshaft, the GS80 displaced 872cc and produced 57bhp. There was also a 55bhp 750cc version entered in the 1979 German 750cc Cross-country Championship. The light duplex frame incorporated a monoshock swingarm and in the 1979 ISDT in Germany (Neunkirchen/Siegerland), Fritz Witzel took out the over 750cc award. The following year the team won the Silver Vase, but all this was a prelude to the most important off-road victories, those in the Paris--Dakar rally.

Billed as the toughest rally in the world, the factory entered three machines in 1981. Prepared by HPN Motorradtechnik, a small tuning firm in southern Bavaria, these had strengthened chassis and long-range fuel tanks. Hubert Auriol rode to an easy victory, repeating this in 1983. This year the engines displaced 870cc, and the suspension included 42mm Marzocchi forks and an Öhlins rear shock absorber. Three-time World Motocross Champion Gaston Rahier joined the team for the 1984 event, winning ahead of Auriol, repeating this on a special 1,000cc version in 1985. Following the death of Thierry Sabine, the rally promoter, during the 1986 rally, BMW disbanded its official works team. HPN continued to develop Paris–Dakar machines for privateers, and a 1,000cc HPN R80 G/S was available in limited quantities through BMW for privateers for 1987. Based on the R80 G/S, but with a Marzocchi fork and twin White Power shock absorbers, the HPN desert racer cost a princely DM29,000. It was still an eminently suitable desert racer, with Munich rider Eddy Hau winning the marathon category for privateers in the 1988 event on a standard HPN R80 G/S.

Above: Rahier on his way to winning the 1985 Paris–Dakar Rally – BMW's final victory in this event for nearly 15 years. *BMW*

Right: The diminutive Gaston Rahier (right) won the 1984 Paris–Dakar Rally from Hubert Auriol (left). *BMW*

Above: The 1995 R80 incorporated many chassis developments from the K-series, but the engine was largely unchanged from earlier models. *Ian Falloon*

Opposite: The Monolever R80RT remained in production for more than ten years, and virtually unchanged from this 1985 version. *Cycle World*

THE R80, R80RT & R65

With the release of the K100, BMW initially decided to end production of the R100 and R80 series. There was already a three-cylinder K75 waiting to fill the mid-displacement class and to BMW's management team the venerable R series seemed an anachronism. The R100, seen at the end of its development life, was discontinued in favour of the K100, and it seemed unlikely that there would still be a market for the smaller twin with its moderate power. But BMW traditionalists saw it differently. To many, no matter how sophisticated and advanced it was, the K100 appeared to be little more than a motorcycle with a small car engine lying on its side. Therefore, in response to loyalist demand, the R80 made a surprising return in 1984. Later, the R100RS and R100RT were also resurrected in response to popular demand.

Although the engine of the R80 was almost identical to that of the previous 1981–84 version, the design of the rocker arm assembly was revised to

(the front now a 19in), and an R65 fork. The kickstart was eliminated and for 1983, there was a revised gearshift providing further improvement to gear selection. Weighing only 198kg (436lb), and with short-travel suspension, BMW created arguably their most competent street bike of the early 1980s. Unfortunately, the market did not see it that way. The high-rise enduro-style exhaust system looked incongruous on a pure street machine, and the 19in wheel appeared to be too small for the R80 G/S frame. While the R80 G/S was in demand, the R80 ST lasted only until 1984. In the meantime, the reputation of the R80 G/S as an unsurpassed all-round motorcycle grew. When the Norwegian Helge Pedersen chose a motorcycle for a ten-year around-the-world trip, it was an R80 G/S. Pedersen covered 350,000km (220,000 miles) on his epic journey.

lower the noise. The rocker arm spindle stanchions were now positively located in the cylinder head, with a plastic washer to eliminate end-play. The result was the side thrust from the rockers no longer pushed the supports apart. Further noise reduction was achieved through rubber buttons between the cylinder head fins. With a new exhaust system that incorporated a pre-silencer box interconnecting the left and right pipes upstream of the mufflers, the noise level was reduced by three decibels while the engine still developed 50bhp.

More changes were incorporated in the chassis. The 38.5mm centre-axle fork was a scaled-down K100-type, with an integral fork brace and a large-diameter hollow axle. Fork travel was less than on earlier boxers, at 175mm. The front brakes were either a single or double 285mm disc with forward-mounted Brembo brake calipers. The wheels were 2.50 x 18in front and rear, and were similar in design to those of the K100, also accepting tubeless tyres. The rear brake was a 200mm Simplex drum, but perhaps the biggest change was to the swingarm and final drive. This final drive, with a bevel roller bearing, was similar to that of the K100, and able to accept a higher load than the previous

needle bearing type. Although the alloy final drive casting looked more massive, it was actually lighter, and carried the lower ratio of the earlier R80/7 (10:32).

The single shock absorber-Monolever rear suspension represented the biggest departure from earlier boxers. Located in a laydown position, the top bolted to the trailing loop of the R80's strengthened main frame, with the bottom located on the final drive housing. There was no provision for damping adjustment and the shock absorber provided 121mm of travel. While the fuel tank was outwardly similar, many electrical components were shifted from the headlight shell to the frame backbone, resulting in a reduced fuel capacity of 22 litres. For 1991, the R80 was replaced by the R100 in the USA, with the 60bhp R100RS and RT engine, although it continued for other markets. From September 1990, all R series twins were available with SAS (secondary air system) as an option. This exhaust emission afterburning system was designed to reduce HC emissions by 30 per cent, and CO emissions by 40 per cent. Already fitted to US and Swiss models, the SAS used exhaust pressure pulses to move two diaphragm valves in the air filter housing, drawing in fresh air. Two tubes directed this fresh air into the

One of several R100RS final
editions was this Rennsport of
1992. *Ian Falloon*

Opposite: Even after the release
of the new R1100RT there was
still demand for the older

R100RT, as the fairing provided
such excellent rider protection.
Ian Falloon

cylinder head and exhaust system behind the exhaust
valve. Development of the R80 was minimal over its
lifespan. For 1990, all boxers received an improved
rear drum brake, the brake pad width was increased to
27.5mm (from 25mm), and there were new mounts for
the brake shoes. There was a floating front disc rotor
from 1991, and Marzocchi forks for 1992.

In addition to the R80, a Monolever R80RT,
replacing the twin-shock version, was also available.
With the same final drive ratio (11:37) the 50bhp
engine still struggled to push the R80RT's considerable

frontal area beyond 160km/h (100mph), but the type
acquired a following amongst those who enjoyed
touring at a more relaxed pace, and required agile
handling. For 1986, a smaller R65, ostensibly identical
to the R80 but with a 48bhp 82 x 61.5mm engine,
replaced the R45, R65, and R65LS. The slight
reduction in power (from 50bhp) allowed the new
engine to run on unleaded fuel, and there was a 27bhp
version for Germany. Production of the two-valve
boxer was envisaged to finish in July 1994, but was
given a reprieve until the end of 1995.

THE R100RS & R100RT

After a two-year hiatus, the R100RS made a surprising return at the 1986 Cologne Show. As with the R80, it was popular demand that saw a new R100RS, and it was originally intended to build only 1,000 examples, but the response was so overwhelming that regular production was instigated for 1988. While the new R100RS looked similar to the 1984 version, there were many differences as it was essentially a larger R80. Inside the engine were the same stronger rocker arm supports as on the R80, and the valve sizes were 42mm and 40mm. Carburation was by smaller, 32mm Bing carburettors, and the exhaust system included a pre-muffler like the R80. This lower state of tune saw the power down to 60bhp, at a moderate 6,500rpm. In a world of escalating Superbike horsepower levels BMW decided to recreate the mould of the R100RS, shifting it from Superbike (as it was in 1977), to that of a more realistic sport-tourer with useable mid-range horsepower. Some enthusiasts were riled that the R100RS was detuned, but it actually had little effect on performance as the new R100RS was as fast, if not faster, than the original. Also, on-the-road performance was superior, especially with the newer and more responsive R80 chassis.

The R80-style frame was similar in design to the pre-1984 version, but with a Monolever swingarm. The wheelbase was slightly less, and the fork rake steeper (27.8°) with 120mm of trail. The wheels, brakes (dual front disc), and suspension also came from the R80, providing the R100RS with lighter steering and more spirited handling than its predecessor. Interchangeable with the earlier R100RS was the fairing, still one of the most effective sport-touring fairings ever offered. There was no longer an optional sport solo seat, and although an oil cooler was standard, the front of the fairing incorporated the earlier open grille.

The R100RS was produced through until 1992, with few changes. There were also several limited edition final R100RS series offered to sustain its model life before the advent of the new R1100RS boxer in 1993, but somehow this didn't achieve the following of the earlier version. Most of the 6,000-odd new series R100RS ended up in Japan, where it was a cult model.

Longer lived was the similar R100RT. Resurrected in 1989, this shared its engine and chassis with the R100RS, but incorporated the larger touring fairing and handlebars of the R80RT. For 1992, it received a Marzocchi front fork and production lasted through until 1996. The final version was the two-tone grey R100RT Classic, initially one of four final 1,000cc 'farewell models' for 1995. Offering a high level of standard equipment, including a custom touring seat and 22-litre top box, the success of the farewell models prompted them to be offered for 1996. But this was their final gasp.

THE R100GS, R80GS & R65GS

While the R80, R100RS and R100RT offered little technological improvement over their predecessors, the R100GS (introduced for 1988) was considerably more than an older R80 G/S with a 1,000cc engine. Although the days of success in the prestigious Paris–Dakar Rally were over for BMW, they were still basking in the glory of their achievements, and this type of dual-purpose machine was still extremely popular in Europe. So good was the R100GS (and its almost identical brother, the R80GS), that it soon became the most popular boxer in the line-up. In order to move from the perception of a combined road/off-road motorcycle, the slash designation disappeared. Now the GS implied off-road sports, and it wasn't an idle claim. Virtually all-new, only the handlebar fittings and headlight were carried over from the previous G/S.

The 980cc and 797cc engines came from their respective road counterparts, although most versions of the R100GS featured 40mm Bing carburettors instead of 32mm ones. The power remained at 60bhp (58bhp with 32mm carburettors) and with the pre-silencer volume beneath the gearbox increasing to 3.8 litres (from 1.5 litres), the torque curve was improved with less noise. With 50bhp, the R80GS engine was identical to the R80's, but with more torque. A different feature from the R80 and R100RS was the layshaft starter (like the K series), weighing 2kg less due to the smaller electric motor but providing the same torque because of the intermediate transmission. The R100GS also included a five-tier oil cooler, which was rubber mounted on the right engine protection bar.

Continuing on from the R80 G/S,
the R100GS was also available as
a Paris–Dakar version from 1990.
Ian Falloon

The chassis was more innovative. The frame was strengthened, with the oval tubes inside the reinforced tank tunnel, with a stronger rear subframe while the Paralever swingarm imparted a new level of handling prowess. This made the GS even more effective on the tarmac than the R100RS and R100RT. Complementing the Paralever swingarm were new 40mm Marzocchi forks providing increased (225mm) travel. Although non-adjustable, the Marzocchi fork incorporated Teflon sleeves, a fork brace, and hollow, 25mm (up from 17mm) axle. Also uprated were the brakes, the front disc increasing to 285mm with a larger Brembo caliper. The 200mm rear drum brake was also lighter,

and operated by a cable instead of a rod. Almost as significant as the Paralever were the cross-spoked wheels. The straight-pull spokes laced from the extreme edges of the rim, with the adjusting nipples in the hub, and permitted the use of a newer-generation Metzeler 'Sahara' tyre.

Although still a reasonably large (210kg) and tall (850mm seat height) motorcycle, BMW obviously realised the strength of the R100 and R80GS lay on the tarmac as there were two mounting positions offered for the front mudguard. There was also a softer spring strut for 1989, and the option of White Power sports suspension for 1990. There was also an improved

THE PARALEVER

Motorcycles, drive shafts, and Munich had been synonymous since 1896 when Hildebrand & Wolfmüller produced a motorcycle with two long con-rods connected to the rear axle doubling as a crankshaft. Then in 1904, the Belgian FN Four went a stage further by incorporating a fully enclosed driveshaft and bevel gears. This system of shaft drive was soon known as cardan-drive, after the Italian scholar Geronimo Cardano (1501–1576) for his design of cardan suspension for compasses. When Max Friz adopted the cardan shaft on the R32, it marked the beginning of BMW's commitment to this drive system, despite its drawbacks and limitations. As suspension travel increased during the 1950s, so did the effects of driveshaft movement on the motorcycle. Becoming more evident with increased power, the R100s in particular rose markedly under acceleration. This was not so much a problem on the road, but off-road riders were less enthusiastic about this characteristic as it used up suspension travel while accelerating. To cancel all forces, the swingarm length needed to be 1,700mm, an impractical figure as this was longer than the wheelbase. A longer swingarm on the Paris–Dakar racing machines helped, as did the Monolever of the R80 G/S, but a more effective solution was required.

During 1983, BMW suspension engineers René Hinsberg and Horst Brenner created the Paralever double-joint swingarm, based on von Falkenhausen's design of 1955 for Walter Zeller's 500cc racer. This design reduced the interference of torque reaction with suspension action and created the effect of a longer swingarm. Inside the single-sided swingarm was a second universal joint, with two additional bevel needle bearings, freeing the rear gearcase and hub assembly and allowing it to float on the rear axle. The gearcase motion was controlled by an alloy strut connecting the bottom of the case to the frame, just below the swingarm pivot. The swingarm, stay arm, gearcase and frame, formed a parallelogram, with pinion torque now feeding into the lower strut rather than the swingarm. The slight fore-and-aft movement of the gearcase was absorbed by the laid down single shock absorber. As the parallelogram arrangement increased the radius of the wheel elevation curve, it provided the same effect as a swingarm 1,400mm in length, and compensation for the acceleration and deceleration forces of 70 per cent, as the optimum wheelbase for a BMW was 1,700mm. This was considered acceptable because too much anti-squat may have appeared incongruous to long-time BMW riders.

Although the weight of the Paralever was 1.6kg (3.5lb) more than the earlier swingarm set-up, it transformed the handling of the boxer. Allowing the full 178mm use of rear wheel travel on the R100GS, the Paralever was so successful that it eventually found its way to the entire range of BMW shaft-drive motorcycles.

Left: The Paralever swingarm included a floating rear gearcase, a second universal joint, and an alloy strut connecting the bottom of the gearcase to the frame. *Ian Falloon*

Below: One of the most successful retro adaptations was the R100R of 1992. *Australian Motorcycle News*

footbrake lever for 1990, along with a Paris–Dakar version. This included additional instrumentation (tachometer and clock), rectangular K75S headlight, and a fairing connected to the enormous 36-litre fuel tank.

While the Paris–Dakar version stretched the off-road association to the hilt, most buyers still thought of the R100GS as an adventure-tourer; a motorcycle ideal for mountain passes and pavements less suited to large touring motorcycles. So, for 1991, the R80 and R100GS became even more street orientated. A new frame-mounted half fairing, resembling that of the previous Paris–Dakar version, provided improved rider protection over the earlier headlight shell/number plate, and included an adjustable windshield. This could tilt back and forth by 75mm. As on the Paris–Dakar, the fairing supports were outside the bodywork, there was a K75S headlight, and inside the fairing was full instrumentation (including a larger tachometer than the

Paris–-Dakar). The front mudguard was low-mounted, and there was a tougher seat with different seat foam. The handlebar switches now came from the K series, with separate buttons on either side for turn signals. Unlike the K series though, they weren't self cancelling.

For 1991, the R80 and R100GS (and Paris–Dakar) also received a floating front disc rotor, and an adjustable Bilstein rear shock absorber with ten positions for rebound damping. Steering head bearing adjustment was changed to the finer thread of the K75 for more precise bearing-play adjustment. The engine was the same as before, but for a 30° exhaust valve seat contact angle to improve seat life on unleaded fuel.

Alongside the R100 and R80GS was a 27bhp R65GS primarily for the German market. Rather than utilising the Paralever swingarm, the R65GS Monolever chassis was identical to the earlier R80 G/S. The engine was from the R65, but building a decade-old model was not really a recipe for success despite its entry level price. By the time production finished in 1992, only 1,727 had been produced. Not since the days of the R50S of 1960–62 had a BMW motorcycle been so unpopular. The R100 and R80GS on the other hand were outstandingly successful, with over 45,000 manufactured. In 1988 and 1989 in Germany, the R100GS was the biggest selling model of any brand.

The R80 and R100GS were available unchanged through until the end of 1994, and for 1995 the 'farewell model' R100GS PD Classic was the only version available. This featured a wider, high-mounted front mudguard, and the older-style round rocker covers of the R100R. As with the other final Classic R100s, production of the R100 GS PD Classic was also extended to the end of 1995.

THE R100R & R80R

By 1992, it was no secret that BMW was developing an all-new generation of boxer twins. As the release of the new model was still a year away, the R100R was created to sustain interest in the boxers. Whereas the R80 and its street derivatives were similar to designs of the early 1980s, the R100R followed the earlier R80ST, and was an attempt to incorporate some of the R100GS developments in a traditional street package. There was no doubt that the R100R was a functionally superior machine to the R80, but for many, the styling was non-cohesive and unsuccessful.

Derived from the successful R100GS rather than the older R80, the R100R was essentially a classic-style roadster. Many components also came from the K series. While the engine was the same 60bhp unit as the R100GS, with 40mm Bing carburettors (32mm for the USA), the classic look was enhanced by a return to the rounder valve covers, introduced on the R68 back in 1952 and last seen in 1976 on the /6 series. Rather than mounting the oil cooler on the engine protection bar like the GS, this was now positioned in front of the engine, and the single round stainless steel muffler was similar to the K series. From the GS came the instruments and 24-litre fuel tank, while the round headlight was from the K75. With an abundance of chrome, and the option of an additional chrome kit, the R100R was a little gaudy for the traditionalist.

The Paralever chassis also came from the GS, but with different wheels, brakes and suspension. For the

first time on a BMW, the suspension was from Japan, Showa supplying the non-adjustable 41mm fork and rear shock absorber. At 135mm, the front fork travel was less than on other boxers, while the rear gas-charged shock provided adjustable rebound damping and 140mm of travel. The front brake was the single 285mm floating disc of the GS, but with a four-piston K series caliper and the option of a second disc. In keeping with the classic theme, the R100R had wire-spoked wheels, a 2.5 x 18in front, and 2.5 x 17in rear. Like the GS, these cross-spoked wheels allowed the fitting of tubeless tyres.

There may have been a few dubious details (like the plastic instrument surround), but the R100R definitely hit the desired mark. It was reasonably light at 218kg (480lb), and with its Paralever swingarm and tighter suspension was the best handling boxer yet. Unlike the ill-fated R80ST, the R100R was seen as a successful adaptation of the GS to the street, particularly in Germany where it became BMW's best selling model since the R45 in 1980. For 1993, there was also a similar R80R, primarily for the German market, replacing the R80. Ostensibly identical to the R100R, but without an oil cooler, the R80R was also available in 27bhp and 34bhp versions for those licence and insurance categories in Germany.

Although the release of the new R259 boxer was creating all the headlines during 1994, the older boxer continued to be extremely popular. The R100R accounted for 22.4 per cent of motorcycle production during 1992, and for 1994, was joined by the R100R Mystic. Designed to present a more classic look than even the R100R, this featured Mystic Red metallic paint, chrome instrument surrounds, lower handlebars, and a more conservatively styled and sporting seat and tail section. As a result, the Mystic was a more successful interpretation of the classic boxer look than the basic R100R. Both versions received twin front disc brakes for 1994, with the SAS emission control system as standard.

'Farewell models' of both the R100R and Mystic were also available for 1995. These were identical to the earlier versions, but for slightly different decals and colours for the seat and various fittings. Production was extended through for the 1996 Model Year and that was the end for the 1,000cc air-cooled boxer. The 800cc engine lasted one more year, in the R80 GS Basic, primarily for the German market. Essentially an

A further development of the
R100R was the more conservative
Mystic of 1994. *Ian Falloon*

earlier R80 GS with a small fuel tank and headlight cowl, the colours were white, with a blue frame, and it featured a Paralever swingarm and the older-style rocker covers.

The final air-cooled boxer, an R80 GS Basic, left the Spandau factory on 19 December 1996. This was the end of an era, but by 1997, the incredible success of the new A60 boxer engine virtually ensured the demise of the two-valve A10 unit. Further competition came from the F650 Funduro single that provided similar power to the smaller air-cooled boxers, in a lighter and more compact package. In the wake of both these newer designs, and struggling to meet increasing noise and emission requirements, the end was inevitable for

the A10 engine. With production lasting 28 years, it was not only the longest lived of all BMW motorcycle engines, but had amassed a huge following of loyal supporters. These engines were astoundingly reliable and refused to die, and a good number of the more than 430,000 produced are still in active service today. Many BMW enthusiasts bemoaned the end of the days of the simple fix-it-yourself motorcycle, but even the most ardent enthusiast could not fail but to be impressed by the new four-valve A60 engine and its innovative chassis. There was no denying the A10 was a great engine, and it powered some wonderful motorcycles, but it wasn't the answer as BMW looked towards the new millennium.

For 1995, the K1100RS was
available as a special model in
Marrakech Red. *BMW*

6 THE K-SERIES

Even when the ground-breaking R100RS was released in 1976, it was evident that the venerable air-cooled boxer engine could not sustain BMW forever. At 980cc and 70bhp, the boxer was at the limit of reliability, but already the market demanded more power. Honda's CB750 of 1969 had rewritten the rules for motorcycles, and like most motorcycle manufacturers, BMW considered a variety of designs between 1970 and 1976. One of these was a 1,000cc air-cooled boxer with overhead camshafts, the K1 of 1976, but the resurrection of the boxer twin would wait. At the end of the 1970s, four cylinders were considered the optimum for higher horsepower with acceptable reliability. BMW needed a different solution to the ubiquitous air-cooled transverse four, but they could no longer follow the flat-four layout they preferred because Honda had 'stolen' that idea with the Gold Wing in 1975.

During 1977, one of BMW's younger design engineers, 28-year-old Josef Fritzenwenger, proposed a longitudinal four-cylinder engine, with horizontal cylinders. Although a longitudinal four-cylinder engine wasn't a new idea, the Belgian FN of 1904 had also featured this design, but with upright cylinders, Fritzenwenger's solution with the engine on its side was unique. Patented under the name of 'The Compact Drive System', initial plans called for the development of two engines; a 1,000cc three-cylinder, and 1,300cc four-cylinder. These two projects, known internally as the K3 and K4, featured engines quite similar in design to those used in BMW's cars, with a single overhead camshaft and rocker valve actuation. The cylinders were inclined to the right, resulting in an extremely large and heavy engine on which it was impossible to install an exhaust system on the same side as the driveshaft.

A reconsideration of the concept led to a pair of smaller engines, a 1,000cc four, and a 750cc triple. With both models able to share the cost of production and development the two-tier concept appealed to BMW Motorrad's new management, Dr Eberhardt C. Sarfert, Dr Wolfgang Aurich, and Karl Gerlinger, who were appointed on 1 January 1979. On 20 February the same year, these directors gave the go ahead for development to proceed. It was a huge step to take. The elegant simplicity of two air-cooled cylinders with pushrod operated overhead valves and twin

carburettors, made way for a liquid-cooled overhead camshaft three and four, with electronic fuel injection.

THE K100, K100RS, K100RT AND K100LT

Under the direction of Martin Probst, the new head of BMW's motorcycle engine development, work began on the four-cylinder engine, known internally as the K589. Probst came directly from working with Paul Rosche on BMW's highly successful Formula 2 four-cylinder racing engines, while Stefan Pachernegg coordinated the design. The first modification was to incline the cylinder heads to the left, and incorporate double overhead camshafts. The engine became known as the A30, and the motorcycle, the K100 (K for Kraftrad, or wheel power).

Almost dimensionally cubical, the horizontal layout provided a low centre of gravity, and exceptional access to the valve gear and crankshaft. The 987cc engine was also remarkably narrow at 504mm, but to keep the engine length to a minimum the bore was 67mm (with a longer, 70mm stroke), with only 74mm spacing between the Scanimet (nickel and carbide)-coated cylinders. As ultimate horsepower wasn't a consideration there were only two valves per cylinder (34mm and 28mm), set at a shallow included angle of 38°, with twin overhead camshafts driven by a single roller chain. The target was only 90bhp, with a broad powerband, and this was comfortably achieved.

To quell the inherent vibration of an inline four-cylinder engine, and the characteristic BMW sideways pitch, the output shaft was positioned underneath the crankshaft, meshing directly and rotating (along with the 460-watt alternator) in the opposite direction at a

Opposite: The K-series not only represented a significant departure from the traditional boxer, but the K1 of 1989 was almost surreal. *BMW*

Below: The K100 development team. From the left: Stefan Pachernegg, Josef Fritzenwenger, Günther Schier, Martin Probst, Richard Heydenreich, and Klaus-Volker Gevert. *BMW*

ratio of 1:1. This secondary shaft carried a torsion damper and incorporated spring-loaded staggered teeth to suppress noise and minimise gear lash. The 180mm dry clutch fed directly from the rear of this secondary shaft, rather than the crankshaft as in the boxer twins. A peculiar characteristic of the K100 was the cloud of smoke which was produced when started following prolonged sitting on the sidestand. Oil seeping past the piston rings would collect in the combustion chamber on the left side of the engine, drawn in by a vacuum if the cylinder was at the compression stroke.

Both the ignition and fuel injection were Bosch, and simplified versions of systems from BMW cars. The LE-Jetronic fuel injection system measured the volume and temperature of the air intake, engine speed and temperature, and throttle valve position. There were individual 34mm throttle valves for each cylinder and the injector-opening period was computed in

STEFAN PACHERNEGG

Born in Graz, Austria in 1943, Stefan Pachernegg, was a member of the Austrian ISDT team between 1966 and 1973, winning two gold medals, and one silver. After working at Bombardier Rotax between 1974 and 1975, he became director of development at Steyr-Daimler-Puch AG Avello in Spain from 1975 to 1979, moving to the company's works in Graz in 1980. He went to BMW as managing director in charge of motorcycle development in 1981, where he initially coordinated the K589 project and initiated the R259 boxer design.

association with the Hall-effect digital trigger ignition.
The ignition advance curve was dual stage, one for
open throttle applications, and another for partial
throttle. The engine layout, along with noise and
emission requirements, dictated liquid-cooling, with the
water pump driven from the front of the output shaft.
The biggest problem with the A30 engine and
transmission layout was the length, and the 'Compact
Drive' system enabled the wheelbase to be restricted to
1,516mm. From the three-shaft five-speed gearbox (in
its separate alloy housing), the driveshaft was secured
directly with a wire snap ring to splines on the gearbox
output shaft. The Monolever swingarm pivoted on
taper roller bearings from the gearbox housing rather
then the frame, and the driveshaft incorporated an
integral rubber shock damper, with the rear drive by
bevel gears. The driveshaft and swingarm were of
necessity as short as possible, and contributed a
noticeable torque reaction.

There were more departures from BMW tradition in
the frame and suspension. The engine and drivetrain
were attached at five points to the tubular steel space
frame as a stressed member. The frame was constructed
of almost straight tubes (30 x 1.5mm and 20 x 2mm).
Weighing 11.3kg (24.9lb), it was an exceptionally rigid
structure, but the engine could not be rubber mounted.
The centre-axle front forks were considerably more
substantial than earlier BMW examples, with 41.4mm
tubes, a rigid 25mm axle, and less travel than before
(185mm). At the rear was a single gas-filled Boge
shock absorber, providing 110mm of travel, and there
were new wheels and brakes. The triple 285mm discs
were slotted, each with a Brembo 38mm twin piston
caliper, while the 2.50 x 18in and 2.75 x 17in cast
alloy wheels were designed for tubeless tyres.
Completing the modern specification was a new
instrument panel with an electronic speedometer which
took its readings from a sensor on the driveshaft. The
sculpted 22-litre fuel tank was made of aluminium.

When the K100 was released in 1983, there were
initially two models: the basic K100 and the faired
K100RS. For the first time on a BMW motorcycle, the
automobile connection was emphasised in the shape of
the radiator grille of the K100. On the K100RS, the
frame-mounted fairing continued the form of the
R100RS, but wind tunnel testing resulted in a smaller
structure, incorporating the mirrors with turn signals,
and an adjustable aerofoil to deflect air over the rider's

helmet. As the fairing increased the top speed from 215
to 220km/h (134–137mph), the K100RS had a slightly
higher final drive ratio.

Although they were still not mainstream
motorcycles, the K100 and K100RS were immediately
successful. The K100RS in particular was arguably the
finest sport touring motorcycle produced up to that
time, and received an enthusiastic reception. Because of
its wider appeal, and despite a six-week metal workers'
strike, production of motorcycles rose to 34,001 during
1984. To accommodate the increase in production, new
manufacturing machinery and robots were installed at
the Spandau works in Berlin.

There were a few improvements made to the K100
for 1985. To combat vibration the front three gearbox
mounting points were changed from rubber-bushed to
solid mount, and the footpeg mounts were shortened,
and mounted solidly. This was intended to shift the
resonant points to make the vibration tolerable. The
1985 version also featured an improved rear disc brake
without slots. The author was one of those converted
to the K100RS during that year. While the unique
character of the twins was absent, and there was still
an annoying high-frequency vibration through the
footpegs at certain touring speeds, the K100RS, with
its optional factory luggage, provided unsurpassed two-
up touring in all conditions. Hard riding could provoke
fading of the rear shock absorber, but the K100RS was
an amazingly competent motorcycle, despite its
considerable dry weight of 249kg (549lb). With its tall
seat height of 825mm, the K100RS was more suited to
larger frame riders. Once the pronounced front-end
dive under braking was mastered through deft control
of the rear brake, and torque reaction controlled
through smooth riding, the K100RS was an
exceptional handler. It provided the author with many
thousands of highly enjoyable touring miles and is
rated as one of the most satisfying, if not the most
satisfying, motorcycles he has owned.

Over the next few years, there were only a few
developments to the K100, many of which were
inherited from the K75. For 1986, there was a
redesigned seat, with grab handles incorporated in the
rear of the seat base, a new footpeg design, and a
mechanical fuel-level sensor. Instead of the previous
two low fuel level warning lights, there was now only
one. Also, there were now fuel tank knee guards for
the K100, and lower fairing air deflectors on the

Opposite: Joining the K-series line-up for 1987 was the most opulent touring BMW ever, the K100LT. All K-series were distinguished by the double-overhead camshaft cylinder head on the left, and the LT came with optional ABS from 1988. *BMW*

K100RS. By 1987, sales of the standard K100 were considerably slower than the sporting and touring variants and it received a facelift. There was now no streamlining, while a lower seat with a 760mm seat height, a smaller K75 fuel tank, and higher handlebars were all fitted. Both the engine and wheels were black with silver highlighting, and the rear brake master cylinder was moved out of sight to under the right side panel. There was also improved fuel tank venting, and from 1990, it came with a digital clock.

There were also developments to the K100RS. A Pearl White motorsport version was available for 1987, featuring blacked-out engine and wheels, and the stiffer, more sporting suspension of the K75S. Using the same outer components, stiffer spring and damping rates provided much less travel (135mm). The fork internals were similar to the K75S, with damping rods in the left leg and a dummy rod serving as a spring guide in the right. Each fork leg carried two full-length springs to provide a two-stage progressive spring rate. The closely spaced coils of the spring were at the bottom, touching the damper piston and ensuring low spring movement. The single gas-charged shock absorber also came from the K75S.

For 1984, a full touring K100RT joined the K-series line-up. With fundamentally the same engine and chassis as the K100, this came with higher handlebars and a larger fairing than the K100RS. The multi-piece fairing rested on vibration dampers and was fastened by a multi-arm support to the handlebar centrepiece. There was also a redesigned two-tiered seat, a wide range of options (including a Nivomat shock absorber), and although the weight of 253kg (558lb) made it the largest ever BMW, when compared with full-dress touring motorcycles from Japan, the K100RT was svelte. Developments for 1985 mirrored those of the other K100s, but included a redesigned windshield and deflectors to flow more air over the engine instead around of the rider's legs.

There was an even more opulent K100LT for 1987. Aimed more at the US market than Europe, this included a Nivomat self-levelling rear shock absorber, factory-installed alarm system, 30Ah battery, engine guards, dual accessory electrical sockets, and a radio installation kit built into the fairing (with speakers, antenna, and wiring, and an optional stereo). With the K100LT, BMW finally began to make some inroads into the full-dress tourer market in the USA, and its success eventually led to larger, and even more luxurious versions. For 1988, it received three accessory power outlets, a larger rear top box capable of holding two crash helmets, and an additional high-mounted instrument panel with coolant temperature gauge and cigarette lighter. There were new wind wings on the windshield.

The final significant development to the K100 was the introduction of ABS on the K100RS Special and K100LT of 1988. These were the first production motorcycles to feature anti-lock brakes as factory equipment, the electronic BMW/FAG-Kugelfischer system being the result of several years' development. The development of ABS for motorcycles followed that for cars, in 1978, with the early adaptation of the automobile system for an R100RS. This was followed by an English hydro-mechanical system on an R90S but results were unfavourable until the incorporation of the ALD anti-lock devise of FAG Kugelfischer in 1983. With this version of ABS, the wheels incorporated 100-tooth pulse generating rings behind each disc. From these rings, inductive sensors sent wheel movement data to an ECU in the tail section under the seat. The ECU compared pre-programmed wheel movement values and when lockup was imminent, a signal was sent to a solenoid/master cylinder pressure modulator mounted above the left footpeg bracket. This reduced braking pressure to the front and rear wheel, and provided a level of safety hitherto unknown to motorcycles. It proved to be a very popular option on the K-series, and by 1992, 50,000 examples, 89 per cent of all the K100-series, were sold with ABS.

There were some other developments for 1988. After five years of trying to overcome smoking start-ups after resting on the side stand, new Citroën-patented pistons and rings solved this problem. Previous attempts through pinning the piston rings, new ring material, and a new cylinder finishing process had failed, but milled cuts and a pinned second ring ensured an air passage through the rings. This prevented a cool-down vacuum. Additionally, the pistons were 15gm (0.5oz) lighter and the K100 engine slightly smoother than before. The final two-valve K-series models were the K100LT and Limited Edition of 1991, with Pearl Green metallic paint, engine spoiler and sports suspension. From May 1991, a retrofitting catalytic converter kit was also available for two-valve models.

THE K75C, K75, K75S AND K75RT

As K100 development progressed, work also began on the A40 (known internally as the K569) three-cylinder 750cc project. From the outset, the 750 was envisaged to complement the larger K100. Providing increased smoothness, with less weight, the K750 was a result of the 'building-brick' principle. This expediency was to reduce development costs, and speed the developmental process. Although essentially three cylinders of the K100, the K75 was to be in a slightly higher state of tune, developing 75bhp (100bhp/litre) at 8,500rpm. Retaining the bore and stroke of the K100, the K75 piston was no longer flat, but designed as a stump cone with valve pockets, which provided an increased compression ratio of 11:1 (as opposed to 10.2:1 for the K100). Contributing to the increased power were shorter intake manifolds and a three-into-one exhaust system incorporating three shorter exhaust pipes. The

valve timing was unchanged while the engine weighed 66kg (145lb), about 10kg (22lb) less than that of the K100 unit.

The three-cylinder crankpins were spaced at 120°, and while this provided perfect balance of the free mass forces, there was a first-order imbalance, similar to that of the boxer twins. Much time was spent overcoming this problem, resulting in two forged balance weights positioned on the output shaft that still rotated at a ratio of 1:1 in the opposite direction to the crankshaft. The rubber damper on the K100 driveshaft to output shaft/clutch connection disintegrated during tests, so the K75 obviously required a larger, more rigid output shaft, without a damper. The damper was shifted to the clutch with a modified smaller and lighter R80-type replacing the K100 clutch. It wasn't until the summer of 1983 that these vibration problems were solved, with the first dynamometer 300-hour endurance tests taking place in October. The Bosch LE injection system was similar to that of the K100, but the VZ-51 L digital ignition was simplified to provide only a single-stage advance curve, which ran somewhere between the two of the K100.

When road testing began in the winter of 1983–84, the three-cylinder engine was placed in a K100 chassis, with re-angled front downtubes to compensate for the shorter engine. Günther Schier, head of Running Gear Development, considered a new shorter frame, but testing confirmed the long K100 frame provided excellent handling, even though the weight distribution was different. With the K75 sharing the basic chassis of the K100, it was then left to designer Klaus-Volker Gevert to provide the K75 with an individual quality. The first designs of 1981 were for the K75S, eventually with a more discreet sporting fairing than the K100RS, the basic K75C design coming a little later. The K75C, like the K100, was planned initially without a fairing, but during 1982 a handlebar-mounted cockpit fairing was incorporated. The frame-mounted K75S fairing

was not designed to provide maximum protection, but to emphasise the more sporting nature of the smaller K-series motorcycle. The wind tunnel-developed nine-piece fairing provided less drag than that of the K100RS (0.45CdA instead of 0.50CdA), as well as a reduction in front and rear wheel lift. For 1986, US models also received a lower cowling which fitted closely underneath the engine, and as this won a prize in the Stuttgart Design Centre, it soon appeared on all versions. With a smaller radiator and fuel tank, the K75 also imparted a feeling of considerably less bulk than its larger brother.

Released for 1986, the K75C shared its 18in wheel,

It seemed a simple procedure to create a 750cc triple out of the K100, but vibration was more problematic than expected. The first of the new K-series was this K75C of 1986. *Cycle World*

twin discs, and front fork (but with new dampers and an integral fork brace) with the K100, while at the rear was a 2.75 x 18in wheel with an integral, rod-operated 200mm drum brake. The rear shock absorber also had a lighter spring to compensate for the lighter weight. For the US market there was a K75T (Touring) model available, with a clear touring windshield and contoured seat. The sporting K75S retained the K100 17in wheel and rear disc brake, but the most significant development was the shorter travel and stiffer front fork (also shared with the 1988 K100RS, and described earlier). As a result, the K75S was a considerably more lithe handling machine than the K100. Not only was the dry weight less at 229kg (505lb), but with reduced weight on the front wheel the steering was lighter. With its smoother engine, the K75 offered a surprising, and cheaper, alternative to the K100, while the K75S was arguably the best-handling BMW motorcycle yet.

One of the problems faced by the K75C and K75S at that time was that the 750cc category was targeted by the Japanese with lightweight high-performance racing-style motorcycles. The Suzuki GSX-R750 and Yamaha FZ750 epitomised this trend, and while the K75 appealed to conservative buyers, it was still more expensive, heavier and less powerful than the Japanese 750s. However, like the K100, the K75 managed to garner a following independent of this fashion-led Japanese incursion, and 20,000 examples were produced in the first three years.

There was an unfaired K75 for 1987, and for 1988,

The K75S was one of the most underrated BMW motorcycles, and was largely unchanged from 1986 until this 1994 version. *Australian Motorcycle News*

the K75S received 30mm wider handlebars, and blacked-out engine and wheels. The seat height of the K75C and K75 was lowered to 760mm and they benefited from a new seat which was without side covers, while the fuel tank gained a wraparound kneepad. The K75 for 1990 also received the sporting K75S suspension, along with the 17in rear wheel and disc brake. New for 1990 was the K75RT, with a similar fairing to the K100RT. This was produced predominately as a police motorcycle, and all K75s this year were available with ABS as an option. The K75S now came with three-spoke K1-style alloy wheels, and with an option of purple paint. The production of the one millionth BMW motorcycle occurred on 18 March 1991. This was a K75RT, donated to the German Red Cross, Berlin, as a first aid accident vehicle.

While the larger K100 was updated and improved during the 1990s, plans to do the same with the K75 were shelved following the release of the new boxer. Consequently, the K75 engine remained almost identical in specification through until its demise, and

the K75 never received the superior Paralever swingarm, but there were a few small developments to the series over the years. For 1992, all K75s were available with a catalytic converter, which was also available as a retrofit option. The K75RT had an optional electrically adjustable windshield for 1993, and all K75s were fitted with a Showa front fork this year. Then, for 1994, the 700-watt three-phase alternator, developed for the new R259 boxer, along with the smaller and lighter 19Ah battery, appeared on the K75.

There were no changes for 1995 as the K75 entered its tenth, and penultimate year, although there was a white (with red seat) final edition K75S. For the final K75 series of 1996, there were two special 'Ultima' editions, the K75 Ultima and K75RT Ultima. With the three-spoke K75S wheels and a few additional features, the K75 bowed out after 11 years with barely a whimper. Production had numbered nearly 68,000, but of all the K-series machines, the K75 was perhaps the least appreciated.

Opposite: From 1988 there
was an optional lower seat for
the base K75, and from 1990,
there was sportier, K75S
suspension and a 17in rear
wheel. *Cycle World*

Below: The K75RT was
visually similar to the R100RT,
and from 1993 had an
electrically adjustable windshield.
Australian Motorcycle News

THE K1

By the end of the 1980s, the Japanese had rewritten the rules for sporting motorcycles. High horsepower engines, excellent handling, and full-coverage fairings were standard fare, but all the Japanese offerings featured chain drive. BMW decided the time was ripe for a shaft drive Superbike, and considerably more performance-focused than the K100. Titled the K1, in the manner of BMW's performance cars (M1, M3, M5, Z1 etc.), this was intended to emulate the classic R90S that had provided unsurpassed performance in its day. Not only would the K1 stand out, but BMW hoped its performance would be class-leading.

When it was first displayed at the Cologne Show at the end of 1988, the K1 certainly shattered the perception of BMW producing conservatively styled touring and sport-touring motorcycles. The styling was dramatically different, and underneath the all-enveloping bodywork was a significantly developed K100. The styling history of the K1 followed a path of

aerodynamic development that began with Henne's record-breaking 500 Kompressor of 1937. Following the release of the K100 in 1983 stylist Karl-Heinz Abe created a sports machine called 'Racer', for the Time Motion exhibition of 1984. This 1:2 scale model inspired the creation of the prototype K1, which was presented to the management on 22 June 1986, when the project was given the go-ahead.

As BMW was committed to the voluntary 100bhp limit for motorcycles sold in Germany, Abe and Klaus Volker Gevert worked at incorporating the aerodynamic advantages of the earlier styling exercise. With its large enveloping, two-piece front mudguard almost mating to the leading edges of the seven-piece fairing through to its large tail with miniature pannier, the CdA factor was a remarkable 0.34, with the rider in a prone position. This drag coefficient was by far the lowest of any production motorcycle. The large tail section incorporated two integrated six-litre storage compartments, and there was a specific 42-litre baggage system available. The bright colours of red or

blue with yellow graphics and highlighting also made the K1 stand out.

There was much more to the K1 than its aerodynamic bodywork. Inside the engine (designated A31), the cylinder head incorporated four valves per cylinder, with two 26.5mm inlet and two 23mm exhaust valves and a central spark plug. This was not exactly ground-breaking technology, and many thought it was surprising that the original K100 hadn't been four-valve, especially considering Martin Probst's racing car background. On the K1, the twin overhead camshafts acted directly on bucket tappets without adjustment shims, with buckets of different thickness used to adjust valve lash. This resulted in a reduction in the mass of reciprocating parts, but required camshaft removal for valve adjustment although the intervals for this were necessarily long, at 30,000km (19,000 miles).

While retaining the undersquare dimensions of 67 x 70mm, there were slightly lighter higher compression pistons (11:1), lighter con-rods, a 1.3kg (2.9lb) lighter crankshaft, and a digital Motronic injection and ignition system. Similar to the systems on BMW cars, there was now no butterfly-type airflow meter, with

airflow being calculated from throttle position and revs. Other sensors determined air and coolant temperature, and atmospheric volume, with an ECU determining ignition timing and injection pulses from a pre-programmed EPROM. While this low resistance digital Motronic injection system contributed around 4–5bhp, unlike the LE Jetronic, it no longer compensated for changes in engine tune. A shorter, cylindrical stainless-steel muffler connected to an expansion chamber underneath the gearbox, and the power was 100bhp at 8,000rpm. There was a slightly higher fifth gear, stiffer clutch springs, and a taller final drive ratio.

Under the direction of Lothar Scheungraber, the chassis of the K1 was developed to provide considerably more sporting performance. Weighing 13.5kg (29.8lb), the frame front downtubes were increased in diameter to 32mm, with 13mm larger rear seat tubes, with the steering head angle reduced to 27°, and with 90mm of trail, the K1 was lighter steering than the K100, despite the wider front tyre. For the first time ever on a BMW motorcycle, the front wheel was a smaller diameter than the rear, the 3.50 x 17in front and 4.50 x 17in rear wheels being wider than before. Manufactured by FPS in Italy, the yellow, three-spoke wheels allowed for the fitment of low-profile radial tyres.

Complementing the wheels and tyres were upgraded suspension and brakes, the K1 using the Paralever swingarm from the GS models, with a single gas-charged Bilstein shock absorber providing 135mm of

Opposite: Following an aerodynamic path begun with the 1937 record-breaking Kompressor, the unique K1 bodywork provided the lowest drag coefficient of any production motorcycle. *BMW*

Below: Underneath the radical bodywork was a completely revamped engine and chassis, with a Paralever swingarm for the first time on a K-series. *BMW*

wheel travel. As the swingarm incorporated an additional U-joint, it was longer than the K100 (450mm), contributing to the long, 1,565mm (62in) wheelbase. The front 41.7mm forks were by Marzocchi, with 135mm of travel, and there was a state-of-the-art Brembo braking system. Larger, twin spirally perforated 305mm front brake discs were gripped by four-piston calipers, with 32mm and 34mm pistons to distribute the pad wear evenly. The floating brake discs were thicker at 5mm and revised lever ratios, with a 20mm master cylinder piston, reduced braking effort. ABS was optional, but standard for some markets like the USA.

The K1 was designed for high-speed use, and the riding position accentuated this. A 780mm seat height, lower and wider handlebars, and higher and wider

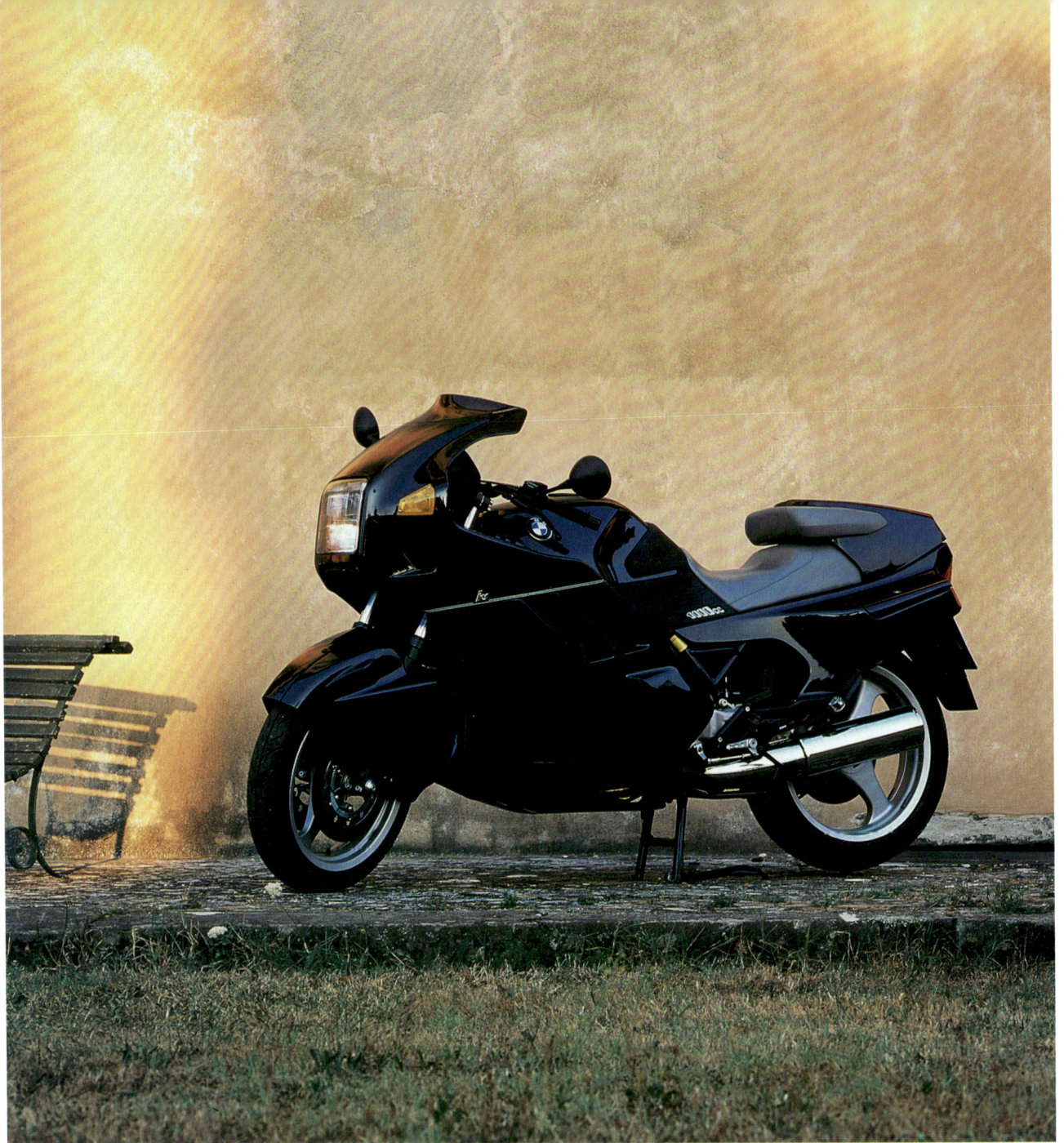

footpegs created an uncompromising sporting riding position. However, even once the lurid colours and unique styling were accepted, the K1 failed in its quest to offer leading Superbike performance. Certainly the faster throttle response, lighter steering, and tighter suspension and brakes made the K1 closer to the Japanese Superbikes than previous K-series, but the dry weight of 234kg (516lb) was intimidating for a sporting motorcycle. The engine, while providing a superb broad powerband with excellent fuel economy, just wasn't powerful enough to overcome this weight obstacle. While the Paralever provided a vast improvement in overcoming the inherent deficiencies of shaft drive for a sporting motorcycle, it still couldn't disguise the considerable unsprung weight. There was

no denying the K1 was the best handling and strongest performing BMW motorcycle to date, but somehow the true nature of the machine was lost in a confusion of purpose.

The first deliveries of the K1 were in May 1989, and it was popular initially, winning many industry awards. Yet, after producing nearly 4,000 during 1989, sales stalled. For 1991, there was a more subdued colour of Classic Black metallic, with silver wheels, and lightened fork compression damping was introduced through increasing the holes in the damper rod from 2.1mm to 2.4mm. This overcame the harshness of the earlier fork when absorbing large jolts at high speed. As extreme engine heat was a problem, there were new baffles to direct more heat out of the fairing slots, but this wasn't

enough to save the K1. By the end of 1993, the K1 was
dead, but today it remains an important representative
of BMW's technological development of the late 1980s.
As less than 7,000 were produced, the K1 is becoming
increasingly sought after as a future classic.

THE K100RS
(FOUR-VALVE)

It was inevitable the four-valve engine and Paralever
chassis would eventually find its way to the K100, and
this happened less than a year after the release of the
K1. Whereas the K1 struggled in its quest to be a
Superbike, the four-valve K100RS made no pretensions
as to its intended function. It also represented a

significant improvement over the previous K100RS,
already the definitive sport-touring motorcycle, but one
that was virtually unchanged from 1983. Not only did
the 100bhp engine with Bosch Motronic MA 2.1
electronic fuel injection come from the K1, but it also
shared the reinforced chassis with Paralever swingarm,
cartridge-style Marzocchi front fork, three-spoke FPS
wheels, and four-piston Brembo brakes. Unlike the K1,
the new K100RS featured rubber front engine mounts
to cut vibration, and the handlebar was wider
(although not as wide as on the K1) for increased
leverage. The rear shock absorber, with less travel
(120mm) came from the K75S, and the front fork
received softer damping and springing than on the K1.
Although the K100RS now shared the higher, K1 fifth
gear, it retained the lower final drive ratio. There were

no changes to the K100RS for 1992 except the rear shock absorber, which was now Showa, and gained infinite rebound damping adjustment.

As the four-valve K100RS shared the excellent fairing of the earlier two-valve version, it functioned similarly. To widen its appeal there was also an optional engine cowling and a lower seat. With improved power, handling, brakes, and minimal driveshaft effect, the K100RS four-valve was just the sport-touring motorcycle traditional BMW enthusiasts were looking for. The K100RS provided the heart of the radical K1 in a more familiar environment, and it proved to be considerably more popular than the K1. From May 1991, a three-way regulated catalytic converter with oxygen probe (reducing hydrocarbons, carbon monoxide, and nitrous oxide) was also offered as an option on four-valve K-series motorcycles with the Motronic injection system. Despite a slight decrease in torque, requiring lower final drive gearing, environmentally friendly BMW buyers accepted the catalytic converter with enthusiasm. By 1992, 41 per cent of all K1 and K100RSs were so equipped.

THE K1100LT AND K1100RS

By 1990, the K100LT was the only K-series motorcycle retaining the two-valve, four-cylinder A30 engine and it was still virtually the same as it was eight years earlier. As had happened with the K100RS, it appeared to be certain that the K100LT would also adopt the four-valve cylinder head and chassis improvements of the K1. The success of Honda's GL1500 Gold Wing proved there was no substitute for displacement with a heavy full-dress tourer, so for 1991 the 987cc engine was boosted to 1,092cc. The displacement increase came through a bore enlarged to 70.5mm, creating the largest capacity BMW motorcycle to date. To reduce vibration the new engine (designated A34) used lighter pistons and 6mm longer connecting rods. With a Bosch Motronic MA 2.2 injection system the power was the same 100bhp as the four-valve K100, but at a lower, 7,500rpm. Most improved was the torque, up to 107Nm at 5,500rpm which was significantly higher

than the two-valve K100 (86Nm at 6,000rpm), and the four-valve K100 (100Nm at 6,750rpm).

The chassis specifications were similar to the K1 and the four-valve K100RS, with a Paralever swingarm, dual front discs with four-piston calipers, and three-spoke wheels. These were not the more modern wider type of the K1, but retained the 18in front and 17in rear of the K100LT (although the rear rim width went up to 3.00in). The rear shock absorber was the Showa unit, shared with the 1992 K100RS. The fairing was moved 30mm forward and incorporated frame-mounted instruments, while the windshield was electrically adjustable. The luggage also provided a larger carrying capacity, and the footrest support plate was remotely mounted to reduce vibration. However, the K1100LT still vibrated, and the rather soggy suspension didn't invite spirited riding.

Opposite: For the tenth anniversary of the K-series in 1994, there was this special edition K1100LT. *BMW*

Below: The K100RS grew to 1,100cc for 1992, and received a new fairing with distinctive side air scoops. *Cycle World*

One year after the release of the K1100LT, the K1100RS replaced the K100RS. The high-torque engine came from the successful touring model, along with the new 700-watt alternator and 19Ah battery of the new generation boxer. There were thicker exhaust headers for increased durability, and the K1100RS featured a modified K100RS chassis. Reinforcement was added to the frame through tiebars connecting the steering head to the rear subframe, and the Marzocchi fork and Showa shock absorber were modified. The fairing received the most development, the trend towards full-coverage bodywork resulting in a new lower section and engine cowl. This incorporated the distinctive BMW 'kidney' grille, along with side air scoops similar to those on a Ferrari Testarossa car. Completing the updates was a four-way adjustable front brake lever and newly designed handgrips. All these developments made the K1100RS the most sporting K-series motorcycle yet. It still had its foibles though, weight and vibration, but as a high-speed long-distance tourer the K1100RS had no peer. Compared with the cheaper Honda ST1100 which was

clawing away at its market share, the K1100RS
provided superior handling, luggage, and overall
capability.

Developments for both the K1100RS and K1100LT
for 1994 included several features from the new R259
boxer. This included not only the new alternator, but
the Motronic MA 2.2 electronic injection and ignition
system, and the option of second-generation ABS II.
The K1100LT received the higher final drive of the
K1100RS, and there was a 10-year K-series anniversary
K1100LT Special Edition. For 1996, the K1100LT
received ABS II as standard equipment, and it
continued largely unchanged for another two years, but
by that time, it really was an anachronism.

As ABS was proving an exceptionally popular
option, the development of the second-generation

ABS II began early in 1990. Again undertaken
in cooperation with FAG Kugelfischer, this retained
the 100-tooth gear and sensor on each wheel,
but incorporated a single, two-channel pressure
modulator instead of two separate modulators as
on the earlier system with three computers
monitoring the signals. The new ABS components
were lighter, weighing 6.9kg (15.2lb) instead of
11.1kg (24.5lb).

There was a special red edition of the K1100RS for
1995, and a black and silver version for 1996, but it
was still out-powered by the new and larger Japanese
sport-tourers, and vibrated too much. The K1100RS
also failed to win the hearts of traditional BMW
buyers. They wanted a new boxer, and the success of
the R1100RS hastened the demise of the K1100RS.

Opposite: Not only did the
K1200RS feature the BMW
Telelever front suspension, but
the aluminium frame was a
radical departure from the earlier
K-series. *BMW*

THE K1200RS

As development resources were fully engaged on the R259 boxer during the early 1990s, it was not until 1993 that the much overdue development of the K-series could begin. Rather than continue along the path initiated by the K1, the development team headed by Jürgen Bachmann decided on a change in direction. Although financial constraints and development time still tied the design to the longitudinal horizontal four-cylinder layout, the new K-series would provide real Superbike performance in a state-of-the-art chassis.

The heavy engine and transmission layout was never going to power a lithe Superbike, so the design parameters were aimed towards a powerful sport-tourer in the best BMW tradition. The first task was to cure the vibration that had plagued the K-series since its inception. Lancaster-type double speed counter-rotating balancers, such as on the Honda Super Blackbird, were investigated, but there was insufficient room inside the existing crankcases for the two shafts and the proposed front Telelever lower arm. Isolating the rider from the chassis by rubber-mounting the handlebars, footpegs, and seat was then considered, but this resulted in imprecise handling. The third solution called for a conventional engine rubber-mounting approach.

Deciding to support the engine in rubber mounts required a new design approach for the frame, so BMW turned to the Italian frame specialist Bimota, in Rimini. Bimota designed a monocoque aluminium backbone-style frame that connected the swingarm to the front suspension mounts. Two 'Silentbloc' bushings, with two lateral steel braces, held the engine in the front, and two rubber/steel mounts were inserted in the frame above the transmission case at the rear. By mid-1994, Bimota's prototype was accepted, and after three versions, the eventual frame used welded cast aluminium components. Four aluminium castings, with honeycomb reinforcing ribs, were welded together to create a hollow backbone frame. Aluminium shanks attached to the rear of this backbone and provided pivots for the rear suspension and mounts for the bolted-on rear subframe. Built by Verlicchi, this frame provided excellent handling, successfully isolated engine vibration, and was easy to manufacture. The only disadvantage was the weight, more than 27kg

(60lb), and 7kg (15lb) heavier than a frame design with a stressed engine.

With the vibration problem solved, the engine could now be developed to produce more power. The voluntary limit of 100bhp in Germany, initiated back in 1978, was now irrelevant and engine development chief Wolfgang Dürheimer coordinated a programme to create the most powerful BMW motorcycle ever. As the 70.5mm bore of the K1100 already left the cylinders with marginal bore spacing, an overbore was out of the question. Stroking was the only way to achieve more displacement, and this required new crankcases, crankshaft, and stronger balance shaft bearings, with a stroke of 75mm providing 1,171cc. This engine was known as the A35.

Although the bore remained the same at 70.5mm, the A35 engine included higher compression (11.5:1), lighter pistons and gudgeon pins, and forged-steel con-rods. The four-valve cylinder head now included 26.5mm inlet and 23mm exhaust valves with thinner stems (reduced from 6mm to 5mm), and with longer duration K1 camshafts, and 38mm throttle bodies, the power was increased significantly, to 130bhp at 8,750rpm. This also saw the maximum revs rise to 9,000rpm, where the piston speed was 22.5m/sec (4,400ft/min). A figure reminiscent of racing engines only two decades earlier, this pushed the crankcase design to the limit so it wasn't surprising the electronic rev limiter cut out at 9,400rpm (which was still up on the 9,000rpm of the K1100RS). There was also an alternative, 98bhp version available, for those who demanded lower insurance premiums. This was essentially identical but for fuel and ignition mapping, and smaller and longer intake manifolds.

Contributing to the power increase was a redesigned intake and injection system. Taking advantage of developments in air-box design, the volume of the air-box increased from 1.0 to 7.0 litres, and was fed by a ram air intake from the front of the fairing. There was also a third-generation Bosch Motronic MA 2.4 engine management system. This had an automatic choke, and it allowed for external diagnostic checking for faults. The alternator was uprated to 720 watts and the exhaust system featured larger diameter manifolds. Because of the horsepower increase, a less restrictive dual catalytic converter set-up was employed rather than a single larger one. Also uprated was the cooling system, with a dual radiator instead of the previous single unit.

The gearbox and clutch were also new for the K1200RS. The smaller diameter clutch (165mm) was hydraulically actuated and connected to a new, six-speed Getrag gearbox. The three-shaft dog-type gearbox was installed in a more compact housing, as it didn't have to support the swingarm pivot, and incorporated an improved torsional damper.

Complementing the more powerful engine was enhanced, sporting orientated suspension. The telescopic forks of the earlier K-series made way for a BMW Telelever system pioneered on the R259 boxer (see Chapter 7). This substituted an A-arm, fork brace, and a ball-joint for the usual telescopic fork lower triple clamp, and differed in detail to the boxer set-up. Marzocchi provided the 35mm forks, while the longitudinal A-arm that braced the lower fork bridge was rectangular in section. Rather than being attached to the engine, this was pivoted on the frame, just above the engine. The ball-joint for the upper fork bridge was contained in the front of the backbone frame and there was a single gas-charged shock absorber. This provided a relatively short, 115mm of wheel travel. The geometry of the A-arm was altered from that of the

R1100RS to provide an anti-dive ratio of 90 per cent (rather than 70 per cent). With 100 per cent providing no dive at all, this greater figure was considered more appropriate for the sporting nature of the K1200RS. A hydraulic steering damper connected the A-arm to the lower Telelever, and the engine was raised 30mm higher than on the K1100RS to allow for an increase in the maximum lean angle from 46° to 50°.

The Paralever swingarm was modified to be mounted on the aluminium frame, and featured revised shock absorber mounting points. As the single shock absorber was mounted at a pronounced angle, it provided a rising rate action, with 150mm of wheel travel. The new chassis provided a slightly shorter wheelbase (1,555mm/61in) than the K1100RS, and more conservative steering geometry (27.25° rake and 124mm of trail). The brakes and wheels were also upgraded for the K1200RS, with the front, 305mm floating discs not now mounted on separate carriers, but directly on the wheel spokes. The five-spoke wheels too were lighter, and wider, with a 5.00 x 17in on the rear. This allowed the fitting of the largest tyre yet on a BMW motorcycle, a 170/60ZR17.

Opposite: With its new bodywork and this distinctive colour scheme, there was no mistaking the K1200RS for its predecessor. *Australian Motorcycle News*

Above: The 2000 Model K1200RS was more subdued in styling, but many found the riding position too sporting in its orientation. *Ian Falloon*

There was also new enveloping bodywork for the K1200RS. David Robb's first major motorcycle design project, he and his 15-strong team set out to create a sporting design that retained an association with the previous K-series. With its slightly more sporting riding position, the seat, and footpegs were adjustable, the K1200RS had a lower fairing, with two-position windscreen, and more modern twin headlight set-up. There were two reflectors in a single housing, with the high beam reflector in front of the low beam. The hand protectors, with their integral turn signals, were intended to retain K-series links, but no longer incorporated rear view mirrors. The large rounded plastic fuel tank and cover seemed excessively wide but the proof of the excellence of the K1200RS was in the riding. Despite the considerable weight of 285kg (628lb),

the K1200RS was possibly the first BMW motorcycle without any peculiar quirks that always set a BMW motorcycle apart. With surprisingly light steering, exceptional stability, no vibration, and a tight driveline, the K1200RS could be ridden incredibly fast with ease.

As the sporting riding position wasn't universally accepted, during 1998 a comfort handlebar and seat were available as an option. This moved the handlebar 40mm closer to the rider, and increased the seat height for both rider and passenger. There was also the option of a wider, 5.50 x 17in, rear wheel, allowing a 180/55ZR17 tyre. By the end of 2000, production numbered more than 21,000, but it was time for a facelift.

Customer demand required improved comfort and weather protection, along the lines of the previous K100RS and K1100RS, along with a less aggressively sporting riding position. For 2001, wind tunnel development resulted in a more slender upper fairing, and a wider and taller windshield. The windshield area was increased by 20 pr cent, and was still two-way manually adjustable. As on the K1200LT, the R1150RT and the R1100S, the two water radiators

were now integrated in a BMW kidney grille in the fairing, with the air scoop feeding air into the intake system opening up at the front right above the radiators. The instrument cover and inner cockpit cover were redesigned, and there were larger rear-view mirrors. The previously optional comfort handlebar was now standard, with both the rider's and passenger footpegs lower. The seat remained with a two-position variable height adjustment, and the rear spring preload featured a hydraulic adjustment. While rider comfort was improved, the biggest development was the incorporation of the new generation Integral ABS. Along with larger, EVO 320mm front disc brakes and new calipers, the K1200RS used a semi-integral ABS with the footbrake only activating the rear disc, and the hand lever acting on all discs. Also new was the option of cruise control. There were no changes for 2002, but the 2003 K1200RS came standard with the previously optional wider rear wheel, direction indicators featuring white glass covers, and there was

Above: For 2001, the K1200RS received a facelift and evolved into a highly capable sports tourer. This is the 2002 version. *BMW*

Opposite: The K1200LT offered a successful new interpretation of the full-dress luxury touring motorcycle. *Australian Motorcycle News*

an option of sports suspension that included stiffer springs and damping.

Although undeniably fast and competent, the K1200RS was still an extremely large and heavy motorcycle in the world of sportsbikes. As its focus shifted towards sports-touring the K1200RS faced increased competition from the Honda ST1300 and the Yamaha FLR1300. Therefore, the next incarnation of the K-series RS was expected to displace nearly 1,300cc, and incorporate some subtle styling alterations to further improve rider protection as the design criteria moved even further away from that of pure sports.

THE K1200LT

Largely undeveloped since its inception back in 1991, the K1100LT was, by 1998, the only remaining Compact Drive System K-series motorcycle. Arguably outclassed by the six-cylinder Honda Gold Wing even when it was released, sales of the K1100LT in the full-dress luxury touring market in the USA were largely insignificant compared with the Gold Wing. The K1100LT was essentially a development of the sport-touring K100RS, and not sufficiently focused and competent to tackle the Gold Wing head on. Like BMW, all the Japanese manufacturers had attempted to break Honda's stranglehold in this lucrative market, but none had been successful. Even as the Gold Wing aged, it remained the standard by which large touring motorcycles were judged, until the advent of the remarkable K1200LT.

Following on from David Robb's K1200RS and R1200C Cruiser, both of which had distinctive new

DAVID ROBB

In 1993, the 37-year-old American-born David Robb was appointed motorcycle chief designer. After graduating from the Pasadena Art Center College in 1979, Robb went to work briefly at Chrysler, and then Audi. In 1984, he moved to BMW, where he was instrumental in implementing computer-aided design for cars. At this time, Robb also developed an interest in motorcycles, and his personal transport of a Kawasaki-engined Bimota KB2 didn't go unnoticed. When Robb went to the motorcycle division of BMW, he not only brought his computer expertise, but also a new approach to design. Rather than maintain the traditional BMW philosophy of continual, and sometimes ad hoc, refinement, Robb advocated an emphasis on creating a unified design that achieved its desired purpose from the outset. He also wanted to create new models that were distinctively different from others in the range, moving away from the previous homogeneous line-up where the entire family appeared closely related. This was to broaden the overall appeal and create greater distinctions between the various models.

personalities and were aimed at a specific rider demographic, the K1200LT was intentionally designed to sit at the extreme end of the spectrum for touring motorcycles. This was unusual, as all previous street BMW motorcycles (and to some extent also the GS versions) were essentially fine graduations on the sport-touring theme. Some were more biased to sport, and others to touring, but sport-touring was the BMW genre and it resulted in some models within the line-up competing directly with others. Thus, the R1100RT boxer, introduced in 1995, was seen as an alternative to the K1100LT, which it was and not a motorcycle with a different focus. This changed with the K1200LT, a serious effort to create a true competitor to the Gold Wing, and one that offered a significantly different interpretation of a touring motorcycle to the R1100RT. There was also a conscious automotive association, with the K1200LT emulating the luxury 7-series executive saloon by providing exceptional comfort, storage space, and luxury.

Although the basic architecture of the K1200LT was closely related to the K1200RS, unlike the earlier LT that was essentially a K100 or K1100RS with extra equipment, the new LT design was unique. The 1,171cc four-cylinder engine was similar to that of the K1200RS, but with a lower compression ratio (10.8:1), longer and smaller intake manifolds, 34mm throttle bodies, and milder camshaft timing. This produced a smoother idle and more consistent off-idle running, essential for low-speed manoeuvrability for such a large motorcycle. The intake system also benefited from a four-chamber Helmholtz resonator that not only improved airflow, but also reduced intake noise. A new stainless-steel exhaust system, with a catalytic converter in the pre-muffler, and a modified Motronic MA 2.4 engine management system, saw the power down from the K1200RS, at 98bhp at 6,750rpm, with the torque 115Nm at only 4,750rpm. Also important for this type of motorcycle was the flat torque curve, with 100Nm available between 2,800rpm and 6,800rpm. As there were many optional electrical accessories, the alternator was uprated to 840 watts.

The clutch and gearbox was different from that on the K1200RS. The hydraulically operated single-plate dry clutch reverted to the larger, 180mm of the previous LT, and the gearbox only featured five speeds, with the fifth gear an overdrive. Like the GL1500 Gold Wing, the K1200LT received an electric reversing

assister, operated by the electric starter motor. Activated only while the engine was running, the switch which reversed the starter motor direction was located above the gearshift lever.

With the intention of producing a large full-dress tourer that provided safe and predictable handling, the chassis was also a development of the K1200RS. The previous K1100LT wasn't known for exceptional handling, as the additional weight taxed the suspension. It was a brave rider who dared push a fully laden K1100LT beyond the threshold of weaves and wobbles, and although the new K1200LT was to weigh a substantial 378kg (833lb), it required sure-footed handling. As engine smoothness was also a priority, the cast-aluminium frame, with rubber-mounted engine, was similar to that of the K1200RS. The steel rear subframe included supports for the integrated luggage.

An advantage of the Telelever was that it was possible to provide front suspension characteristics suitable for a specific purpose. The anti-dive ratio could be altered through the geometry of the A-arm, as could the amount of spring travel while maintaining damping control. This feature was particularly appropriate on a large luxury touring motorcycle that required soft and supple suspension with positive wheel control. The front shock absorber provided 102mm of spring travel, and while the huge, tiller-style, handlebars were rubber mounted as on the R1100RS, they had separate bearings to neutralise the Telelever sway forces.

Although the K1200RS had a long wheelbase, the K1200LT was stretched out even further to provide more space for the rider and passenger. Achieved through an 80mm longer Paralever swingarm, the rear shock absorber spring travel was 130mm, with hydraulic spring preload adjustment. The front 17in wheel and 305mm brakes came from the K1200RS, as did the rear 17in wheel, but the rear disc was increased to 285mm, with a four-piston caliper. ABS II was standard.

Intended to convey an aura of luxury, comfort, and convenience, the integrated bodywork was quite unlike earlier BMW motorcycles, and was a new concept in motorcycle design. The fairing, fuel tank, seats, side luggage, top-box, and even the exhaust system were integral parts of the complete motorcycle body. The cockpit was almost automotive-like, the wide fairing offering unparalleled wind and weather protection,

with the built-in, non-removable luggage compartments flowing out of the slowing side panels. Tipping-over protection was sensibly provided through impact strips at the rear and a bumper on the sides of the fairing, with the tipping forces feeding directly to the frame in the style of older crash bars.

Comfort levels were predictably high, with BMW engineers thinking of nearly everything. A thumb switch on the handlebar adjusted the windshield height, and there were even two sizes of windshield available. Transparent deflectors on the sides of the windshield protected the rider's hands, with additional wind deflectors beneath the rear-view mirrors. Easy adjustment of the rider's seat was provided through a lever under the seat, while a gas damper kept the seat open. The seat could be set at 770mm or 800mm, with the passenger backrest adjustable by 25mm through moving the top-box. A sophisticated locking system on the two, 35-litre side-storage compartments allowed them to open at the pull of a lever, with a gear inside the compartment cover activating a consistent seal so there was no need to lock them while riding.

Unlike the K1100LT, it was clear the K1200LT's luxury features were intentionally integrated into the design from the outset. No longer was the radio an afterthought installed in the fairing glove box. Now, the sound system included hand-operated controls, with separate switches at the rear for the passenger, four speakers, and the option of a six-CD changer. Other options extended to heated seats, a trip computer, cruise control, intercom, and a mobile phone. The most extraordinary option was a dealer-fitted refrigerator installed in the top-box. There were additional chrome kits available, and for the 2000 Model Year the top-box incorporated an additional brake light.

To many, the K1200LT may have seemed incongruous in the predominantly sport-touring line-up, but it provided BMW with credibility in the full-dress tourer market for the first time. It may have been extraordinarily heavy, but weight wasn't considered a handicap in this sector. BMW set out to build a better motorcycle than the GL1500 Gold Wing, and they succeeded. The K1200LT was lighter, had superior suspension, a considerably more efficient engine management system, and a higher level of equipment. Despite giving away 300cc, the engine was much more economical, and performance was comparable. The

The third generation of ABS, the
BMW Integral ABS system, was
first offered on the K1200LT from
2001. *BMW*

press enthused over the LT, and in the market where it
was intended, the USA, it immediately jumped the
R1200C cruiser to become the best-selling BMW
motorcycle in 1999. In the USA, it was available in
three equipment levels, the Standard, Icon, and
Custom, and by the end of 2000 production numbered
around 15,000 units.

The big news for the 2001 Model Year was the
introduction of the third-generation ABS on the
K1200LT. Known as BMW Integral ABS, and developed
in conjunction with FTE Automotive in the Bavarian
town of Ebern, this system was initially only available
on the K1200LT, but was to be installed progressively
on new models in the future. Even lighter than ABS II,
weighing only 4.36kg (9.6lb) including sensor rings

and wheel sensors, Integral ABS provided many more
functions than the earlier system. An electro-hydraulic
brake servo for each wheel built up brake pressure
more quickly and equal brake distribution was
provided to both the front and rear wheel brakes
through either the foot pedal or hand lever. The system
provided faster ABS response, and the larger (320mm)
front discs produced 20 per cent more stopping power.
All the control components were located in one compact
housing, the so-called pressure modulator. Also new for
the K1200LT for 2001 was optional GPS navigation,
the first time this was available for a motorcycle, the
map screen being located on top of the dummy fuel
tank with control from the handlebars. Like the
K1200RS, cruise control was available for 2002.

New for 2003 was the K1200GT,
a grand tourer based on the
sporting K1200RS. *Ian Falloon*

THE K1200GT

Supplementing the full-touring K1200LT, but providing a more touring emphasis than the K1200RS, was the K1200GT (Gran Turismo or Grand Touring) for 2003. Another example of the new BMW design philosophy of focusing on specific categories, the K1200GT was intended to plug the wide gap between the luxurious LT and the almost-Superbike sporting RS. This was the motorcycle for fans of high horsepower four-cylinder engines, but who required touring comfort. In many ways the K1200GT continued where the K100RT left off back in 1989. While the subsequent K100LT, K1100LT, and K1200LT expanded the luxury quotient, becoming increasingly opulent in the process, all this equipment sacrificed ultimate performance. The K1200GT still offered the essential touring equipment, but at 300kg (660lb) was not excessively heavy for this type of motorcycle, and was only marginally heavier than the R1150RT boxer.

The engine and drivetrain of the K1200GT was identical to the K1200RS, all K-models for 2003 receiving a modified oxygen sensor, catalyst coating, and engine management, to further reduce emissions. New for the K1200GT was the fairing, now wider at the bottom, and the electrically adjustable taller windshield. Complementing the taller fairing were higher handlebars, which provided a more upright riding position, and a new two-way height-adjustable seat. Standard equipment on the GT included partially integrated ABS, and a luggage rack and cases to match the bodywork. As with the K1200RS, the range of options extended to a heated, lower seat, and cruise control system.

Optional on the R1100RS was
an integrated full fairing that hid
most of the engine. This 1997
version included a standard
catalytic converter. *Australian
Motorcycle News*

7 THE NEW BOXER

Although the K-series of 1983 was designed to supplant, and eventually replace the boxer twins, this did not actually happen. Public demand for the R-series boxer was such that production not only continued alongside the K-series, but the boxer's share of BMW motorcycle production remained consistently between 40 and 50 per cent throughout the 1980s. It was obvious that the boxer was not about to die, and as early as 1984, plans were considered for a modern boxer. This 800cc boxer was initially envisaged to supplement the K-series, and was positioned in the mid-range market. By the time development was approved, market considerations dictated 1,000cc, and the basic layout of a longitudinal twin with the cylinders protruding into the airstream was a required design parameter. It was important to BMW that the boxer tradition was maintained, but the new engine needed to be more powerful, and narrower, to provide increased ground clearance with the higher lean angles possible with modern tyres.

Under the guidance of Stefan Pachernegg, a 'Boxer Workshop' was instigated. Increasing noise and emission regulations demanded four-valve cylinder heads, after testing with both three-valve and five-valve layouts showed them to be unsatisfactory. A conventional double overhead camshaft layout was also unsuitable because of the requirement to limit engine width. This led to the consideration of several valve train designs during 1985, and in April 1986 the first design model was presented to the management of BMW Motorrad. At this stage the valve layout was still undecided, and it was not until Georg Emmersberger produced a hybrid layout during 1986 that a solution was found. Emmersberger's design featured an intermediate shaft beneath the crankshaft, driving two roller chains to a single camshaft in each cylinder head. To minimise engine width, the camshafts were below the four valves, and were actuated by rockers through short pushrods. By April 1987, the prototype engine with Emmersberger's valve layout was complete, and the development was authorised under the designation R259. The project manager was Richard Kramhöller, and what was eventually a 1,100cc engine, became known as the A60 as 1,000cc and A61 as 1,100cc. From the outset, it was also agreed that the engine structure would be load-bearing, and even before the engine's valve train system was

One of the most successful series
in the history of BMW
motorcycles has been the RT, and
undoubtedly one of the finest of
these is the R1150RT. *Ian Falloon*

The R259 project did not only
include a new boxer engine, but
the chassis with its Telelever front
suspension was equally
innovative. *BMW*

finalised, work was progressing on a new front
suspension system.

Seven development 1,000cc engines were prepared,
the first running in February 1988. Although it
immediately produced an impressive 84bhp, there were
problems with valve control over 7,000rpm, the
rockers not following the camshaft. The first step in
solving this problem was to mount the camshaft, with
the rockers, in a separate sub assembly within the
cylinder head. Successive rocker designs culminated in
chill-cast tappets operated by the camshaft, with steel
ball heads on the other end of the short aluminium
pushrods. This prototype engine was tested in May
1988, and presented to the management in October.
The project received the go-ahead for further technical
development, but not for the styling. This was to

involve the car stylists from sister companies BMW AG
and BMW Technik.

It was not only the engine design that was
innovative – even more so was the chassis, and in
particular, the suspension. BMW had a long tradition
of following a path in motorcycle suspension design
different from other manufacturers. From the first
production hydraulic telescopic forks of 1935 to the
Earles fork of 1955, they had shown they were not
afraid to break with tradition. The Earles fork, with its
excessive unsprung weight and high steering inertia,
was no longer considered suitable, but BMW
considered the telescopic fork overworked. It had to
perform four functions; guiding the wheel, springing
and damping, steering, and support under braking. For
the new boxer, BMW decided to develop a new front

suspension set-up, one combining the telescopic fork with support arms resting on the frame and serving as a separate, independent suspension.

First patented in Britain in the early 1970s by Laurence King and John Pizzey, with the support arms linked to the bottom of the telescopic fork, it was not until 1981 that the system was considered by BMW. The British engineer Hugh Nicol, sent BMW details of his Nicol Link Suspension System which combined the telescopic fork with a longitudinal arm linking the fork bridge to the frame. At around the same time, Phil Todd and Nigel Hill produced the Motodd Laverda with an identical suspension layout to Nicols's. When this received positive test reports in 1984 BMW's engineers decided to reconsider the design, concluding that it provided the ideal solution for front-wheel control. With the self-supporting engine block, it was possible to mount the control arm on the engine, directly transmitting the brake forces and providing an anti-dive effect without additional components. Further development led to a longitudinal A-arm connecting the telescopic fork and the engine, with ball-joints between the arm and the lower fork bridge, upper fork bridge and the steering head tube. A single shock absorber connected the A-arm to the frame.

THE R1100RS

The first model to feature the new A61 engine and A-arm fork was the flagship model, the sport-touring R1100RS. The initial designs for the R1100RS were produced during 1989, alongside the development of the engine and suspension. At this stage, the 1,000cc engine was still fuelled by carburettors, and incorporated an oil bath alternator running from the crankshaft on the front of the engine. This was subsequently moved on top of the engine, driven by a V-belt. Testing revealed noise and vibration problems, poor performance and crankcase cracking, and the engine therefore required considerable redesign. Dr Burkhard Göschel, the new Head of Motorcycle Development, decided to increase the capacity to 1,100cc, and concentrate exclusively on development with Motronic injection. The power remained at 90bhp, but with increased torque and with a maximum lean angle of 49° provided by the narrower engine this ensured the R1100RS would prove to be a competent sporting motorcycle.

By the second half of 1990, three developmental motorcycles were successfully completing endurance tests, and by early 1991 there were new crankshafts and a modified valve control system. The engines were now proving reliable, but the front suspension required further development. Lothar Scheungraber headed the running gear development team, and longer and thinner front sliding tubes, a shorter and less-rigid control arm, and less brake dive compensation, all contributed to the fine tuning of the front suspension. In 1992, this became known as the BMW Telelever suspension. The drive development team of Christoph Schausberger also modified the rear Paralever, with the suspension strut positioned in the middle of the round-section swingarm.

The R1100RS styling initially presented many problems for the team headed by Klaus-Volker Gevert and Markus Poschner. The fairing and fuel tank were required to provide ample protection, streamlining, ergonomics, and sufficient fuel capacity. However, the large alternator and front subframe made it difficult to incorporate a conventional fuel tank. The final solution involved testing in the BMW wind tunnel in South Africa, with the plastic, 23-litre fuel tank enveloping the sides of the motorcycle.

When it appeared in its final form at the Cologne Show at the end of 1992, the R259 represented a continuation of Max Friz's nearly 70-year-old boxer concept, while incorporating many new developments. Instead of the long-serving one-piece tunnel housing the crankcase that presented a number of manufacturing difficulties, there was now a pressure die-cast vertically split two-piece crankcase. This incorporated the oil sump and was necessarily strong as it formed a load-bearing member of the chassis. The four-valve cylinder head featured two 36mm inlet, and two 31mm exhaust valves, with a single, centrally positioned spark plug, and the cylinder bores were coated with Gilnisil, a nickel-silicon layer by the Italian manufacturer Giladoni.

Despite their larger diameter, the 99mm pistons provided a 10.7:1 compression ratio, and weighed a third less than those of the older boxer. While the con-rods were made of sintered and forged steel, these were more consistent in weight and featured the crack technology used on BMW cars. The big-end of the con-rod was intentionally fractured rather than sawn, and the advantages were the larger common surface when

The team responsible for the R259 project. From the left: Wolfgang Dürheimer, Klaus-Volker Gevert, Markus Poschner, Burkhard Göschel, Jürgen Kurzhals, Christoph Schausberger, Richard Kramhöller and Lothar Scheungraber. *BMW*

bolted together. The one-piece crankshaft provided a marginally shorter stroke than the earlier boxer (70.5mm), also using only two bearings. A centre bearing would have necessitated additional cylinder offset, but the rear bearing was a double journal type that didn't require thrust washers to set the end-play during assembly. With a displacement of 1,085cc, the R259 was the largest BMW boxer engine yet.

One of the biggest developments was the incorporation of two oil circulation systems, one for lubrication, and one for cooling. Two Eaton pumps were housed in a separate unit on the front of the intermediate shaft, the cooling pump at the front and the lubrication pump at the rear. The cooling oil pump was a volume type, designed to circulate as much oil as possible. With oil travelling to the cylinder heads, and

around the exhaust valves from the oil cooler to the sump, the engine became known as the 'Oil-head' in the USA (as opposed to the older 'Air-head'). There was also a sophisticated crankcase ventilation system, including an oil separator.

The intake system for the R1100RS reflected all the recent advances in technology with this development being coordinated by Jürgen Kurzhals. The air was drawn into a large air-box through carefully designed intakes and a snorkel beneath the fuel tank. One of the advantages of electronic fuel injection, especially for twin-cylinder motorcycles, is that large-diameter venturis can work acceptably over the entire rev range. The exhaust headers were 38mm in diameter, and with a Bosch Motronic MA 2.2 system, the R259 engine produced 90bhp at a moderate 7,250rpm. Torque was

With its high camshafts and short pushrods, the A61 engine provided nearly all the advantages of a double-overhead camshaft design, but in a narrower layout. *BMW*

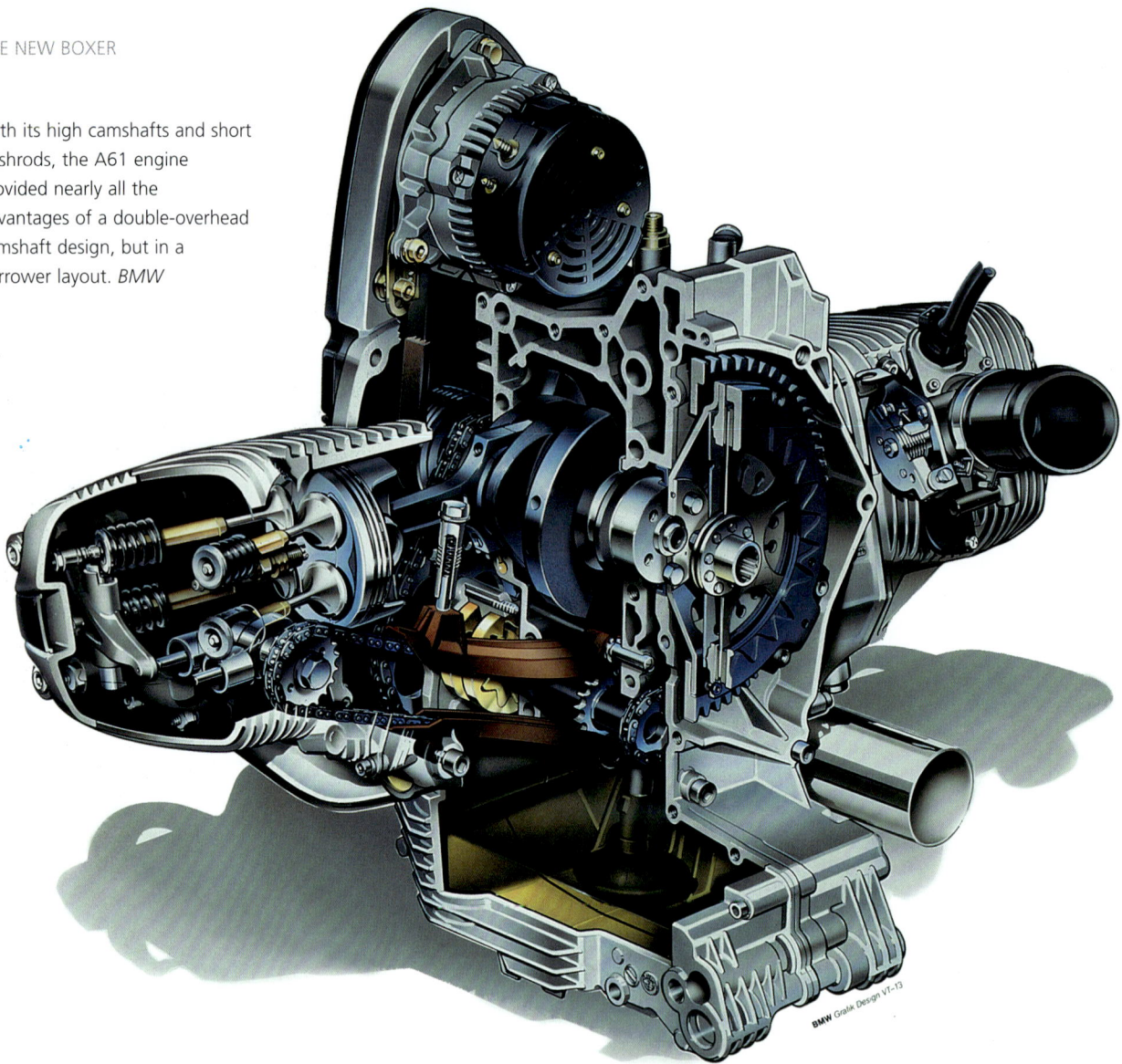

an impressive 95Nm at 5,500rpm. As with the earlier Motronic MA 2.1, several sensors determined the engine's current operating conditions and compared this data with that of an EPROM in the CPU. A catalytic converter was also available, this incorporating an oxygen sensor upstream (rather than downstream like the K1100). Although the drivetrain, including single-plate dry clutch, K-series three-shaft five-speed gearbox, and Paralever double U-jointed driveshaft was similar to other BMW motorcycles, the electrical system was upgraded considerably. A three-phase 700-watt alternator ensured reliable starting, and allowed for a smaller and lighter 19Ah battery.

The engine may have continued the boxer tradition, but there was nothing traditional about the chassis of the R259. Not only was there a new front suspension design, the BMW Telelever, but there was no conventional frame as such. The engine unit supported both the front aluminium and rear steel subframes, along with the front Telelever and Paralever rear swingarm. The production form of the BMW Telelever

consisted of a 35mm fork, with 620mm tubes for increased overlap. The tubes were oval section with Teflon-coated bushings, but did not contain springs or dampers. The fork was thus only for wheel location and steering, with most of the braking forces absorbed by the A-arm, with springing and damping by the central, non-adjustable Showa shock absorber. The Telelever provided 120mm of front wheel travel, and allowed for excellent weight distribution, with 52.7 per cent of weight on the front wheel. The steering head angle was a steep, 24.1°, with only 104mm of trail. At the rear, the Paralever swingarm was 520mm long, the Showa shock absorber providing infinite damping and 135mm of travel. The wheels and brakes were identical to those of the K1 and K1100RS, twin 305 x 5mm floating cross-drilled discs on the front with four-piston Brembo calipers (32mm and 34mm pistons), and a single, 285 x 5mm disc with a dual, 38mm piston Brembo caliper on the rear. The second-generation ABS II was an option.

Although it was no lightweight, at 239kg (527lb),

the R1100RS was an eminent successor to the much-loved R100RS. The styling, while undoubtedly aerodynamically excellent was not to everyone's taste however. Not classically beautiful in the mould of the R90S or R100RS, to some observers it appeared like a large grasshopper, especially from the front. Another unloved feature was the rubber-mounted handlebar that followed the Telelever movement and provided a feeling of vagueness. Although the gearbox was often considered to be unrefined, there was still no denying the overall excellence and competence of the R1100RS. BMW had gambled considerably with the R259, and they had got it right. Here was a BMW motorcycle that appealed to both the traditional enthusiast and the modern rider, and the new boxer

The R1100RS was a sport-touring motorcycle par excellence, providing adequate power with superb handling. *Australian Motorcycle News*

soon overtook the K1100RS as the premier BMW sports-tourer. Despite its intensive development, the first year was not without teething problems. There were cases of oil leaks caused through poor crankcase ventilation, and some gearbox problems also. These were rectified the following year, with O-rings between the gears and their shafts, and tighter tolerances on the input shaft.

Standard for 1994 (in Germany and some selected markets) was the previously optional ergonomic kit that included an adjustable windscreen and handlebars, a variable seat height, and rider information display. The front Showa shock absorber now included more compression damping and a stiffer front spring. Many suspension and engine components were available in black. The ergonomic kit was standard worldwide for 1995, as was a catalytic converter for 1996. This year the front mudguard extended rearward with a component from the new R1100RT, and the inside of the fairing was no longer black, but the same colour as the full fairing.

From 1997, the front shock absorber received

RETURN OF THE SINGLE: THE F650

Looking for an entry-level machine to complement the R259, BMW decided to follow a different path to that pursued back in 1978 with the R45 and R65. The smaller twins were almost as expensive to produce as their larger brothers and profitability was marginal. A smaller displacement version of the R259 was thus discounted, and in the interests of developing a motorcycle as quickly as possible BMW considered a joint project with another manufacturer.

During 1990, BMW noticed with interest, the development of the Aprilia Pegaso 650, and project E169 was approved in June 1991. The design for a light, enduro-style motorcycle began immediately. Following the approval of Bavarian-based British designer Martin Longmore's styling proposal, a full-scale model was completed by January 1992. The first road-going prototype was completed by June that year, and this led to the signing of the three-way joint-venture contract with Aprilia and Rotax. Aprilia, already a major motorcycle manufacturer, producing over 50,000 motorcycles per annum, would assemble the new model at their plant at Noale in Northern Italy. The Austrian engine manufacturer Rotax, would supply engines similar to that used on the Pegaso. Development proceeded extremely quickly, with the first production versions rolling out of Noale in September 1993.

Various changes were made to the existing liquid-cooled Rotax Pegaso engine. The crankshaft bearings were plain instead of roller, and there was a four-valve, rather than five-valve, cylinder head. The valve sizes were 36mm and 31mm, and the double overhead camshaft set-up with bucket tappets was similar to the K1. A roller chain on the left drove the camshafts. Carburation was by two Mikuni 33mm CV carburettors, and there were two spark plugs per cylinder.

The bore and stroke were 100 x 83mm, giving a displacement of 652cc, and there was an engine-speed gear-driven balance shaft in front of the crankshaft. The power of the A80 engine was 48bhp at 6,500rpm, with a lower power, 34bhp version available for the two-tier European licence regulations. Lubrication was dry sump, with the oil tank in the upper part of the frame, and in a departure from usual BMW practice, the five-speed gearbox was incorporated in the crankcases. Perhaps the most significant departure though was the final drive, which was an O-ring chain. The E169 was the first BMW motorcycle not to feature shaft final drive.

Unlike the Aprilia Pegaso, BMW chose to include a steel rather than an aluminium frame. Single-loop, this consisted of sheet and square-section steel, with the engine serving as a semi-stressed member. The steering geometry of a 28° steering head angle and 116mm of trail, indicated an emphasis on road rather than off-road use. The swingarm was a twin-sided deltabox type, with a rising-rate linkage to a single Showa shock absorber and 165mm of rear wheel travel. Also from Showa came the 41mm telescopic fork (with 170mm of travel), the 300mm front and 240mm rear disc brakes were Brembo, and the 19in front and 17in rear wheels wire spoked. Weighing in at 189kg (417lb), the F650 was marketed as a 'Funduro' and soon set new standards for middleweight, dual-purpose machines. Its off-road capability may have been marginal, but as in the tradition of the classic R80 G/S, the F650 was one of the most competent handling tarmac motorcycles available. Not surprisingly, it was an immediate success, and by July 1994, 10,000 had been produced.

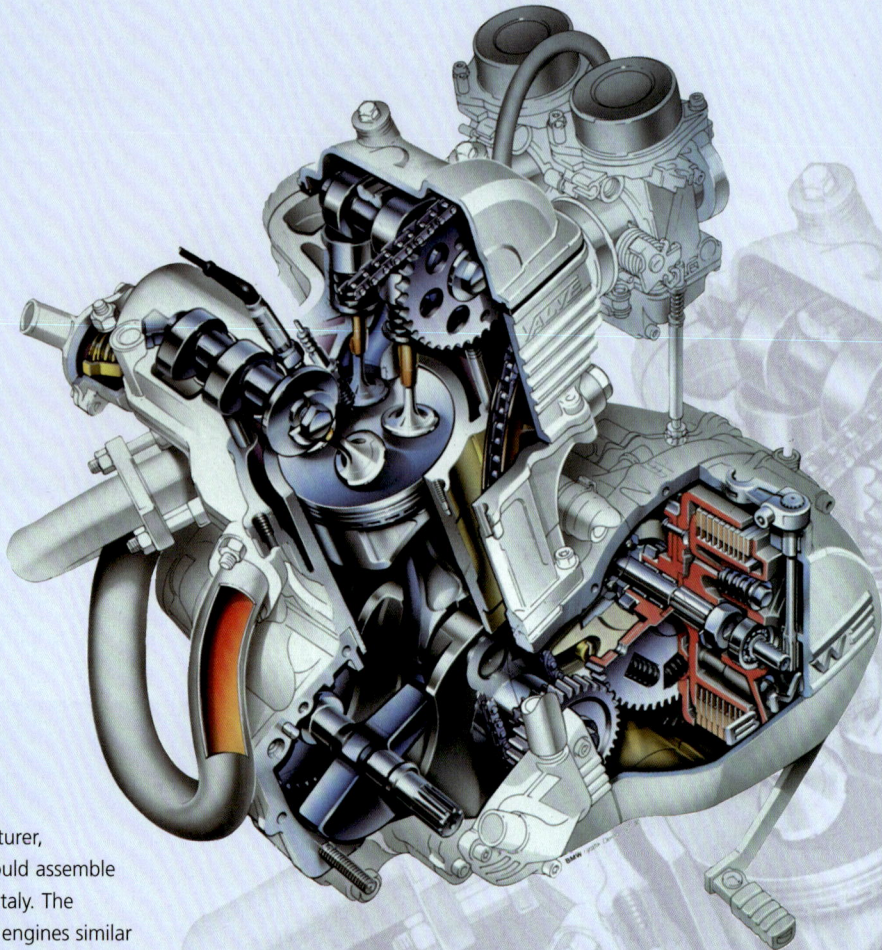

There were minimal developments to the F650 over the next few years, but standard for 1995 was a catalytic converter, and there was a centrestand (previously optional) for 1996. With the demise of the traditional air-cooled boxer during 1996, the F650 range was expanded for 1997. Not only did the Funduro receive a

Opposite: The A80 engine for the F650 was built by Bombardier-Rotax to BMW requirements. There was a four-valve cylinder head and twin Mikuni carburettors. *BMW*

Right: Although not a true off-road motorcycle, the F650 Funduro was an excellent back road machine. *Australian Motorcycle News*

Right: After a facelift for 1997, the final F650 left the Noale production line at the end of 1999. *Australian Motorcycle News*

facelift, the F650ST 'Strada' joined it as a pure road bike, primarily for city use. There was now a taller fairing and windshield, a clock, and lower seat for the Funduro, while the F650ST featured an 18in front wheel, a smaller fairing, and an even lower seat (785mm). This could be lowered further to 735mm with an optional suspension kit. The ST also had more street-orientated steering geometry, and a shorter wheelbase.

A special version of the F650 was launched early in 1999,

followed by a similar F650ST. These were provided with additional equipment, the ST receiving heated handlebars, higher windscreen, a rear top-box, and a catalytic converter. Production of the F650 and F650ST by Aprilia at Noale concluded at the end of 1999 following the termination of the assembly contract. With more than 64,000 units manufactured, production of the F650 and F650ST had exceeded all expectations, and a new assembly plant was then built at Berlin-Spandau.

adjustable rebound damping through a slotted screw, with hydraulic adjustment of the rear shock spring preload, and for 1998 there was a special 75th Anniversary edition. The final R1100RS of 2002 received larger front discs (320mm) and EVO brake calipers. By the time it was replaced by the R1150RS, the R1100RS was the longest-serving boxer in the line-up, but it never proved to be as popular as the GS and RT variants.

THE R1100GS & R850GS

The second new-generation boxer was the R1100GS, introduced for 1994. Given the popularity of the GS-series this was no surprise. From 1980 until mid-1993, sales of G/S and GS models numbered 62,000, making this series one of BMW's most successful. The R1100GS was designed to continue the tradition of the spirit of adventure that began with the first R80 G/S, although it was considerably larger (both in size and displacement), and more complicated. As with earlier GSs, the R1100GS was derived from its street relation, in this case the R1100RS, but unlike earlier versions, the engine was developed to improve off-road performance. Tuned to produce less power, but more torque, the A61 engine received different camshafts, modified exhaust manifolds with stainless steel mufflers, recalibrated Motronic injection, and a slight reduction in the compression ratio (to 10.3:1). This translated into 80bhp at 6,750rpm, with torque up slightly to 97Nm at 5,250rpm. A lower final drive ratio ensured more sprightly top gear roll-on performance than the R1100RS.

While retaining the three-piece frame layout of the R1100RS, there were a number of developments to provide dual-purpose performance. The front suspension travel was increased to 190mm, requiring a longer stroke shock absorber and the A-arm and front subframe were modified resulting in a steeper steering head angle (26°), a longer wheelbase (1,509mm/59.5in), and 115mm (4.5in) of trail. The different geometry of the A-arm also resulted in a stronger 90 per cent anti-dive ratio. The rear shock absorber incorporated a hydraulic spring preload adjuster, the longer travel unit providing 200mm of wheel travel. Another important difference to the R1100RS was the

handlebar mounts, the extremely wide (820mm) handlebar now mounted rigidly on the fork bridge, and connected to the Telelever fixed-position tubes through two ball-joints. This isolated the handlebar from the tilt and swivel motion of the Telelever, that was evident on the R1100RS.

The wheels were the familiar cross-spoke style, a 2.5 x 19in on the front and a wide, 4.0 x 17in on the rear. Also familiar were the Brembo brakes, although there was a smaller, 276mm, disc on the rear. ABS II was also available as an option, but due to the desirability of being able to lock the wheel in gravel this could be deactivated manually. Completing the level of equipment was a 25-litre plastic fuel tank, small cockpit fairing with adjustable windscreen, and no less than four mudguards. At 243kg (535lb), the R1100GS was certainly no off-road motorcycle for the faint-hearted. The size too was intimidating for a smaller rider, and if the styling of the R1100RS hadn't met with universal acclaim, that of the R1100GS was almost contentious. There was certainly no mistaking the R1100GS for any other motorcycle. With its upper 'beak-like' front mudguard, channelling air into the oil cooler under the headlight, the R1100GS may not have been beautiful, but it was undeniably purposeful. Here was a motorcycle that successfully continued the GS tradition. Also more suited to the tarmac than the dirt, the R1100GS, with its wide handlebars, unlimited ground clearance, and supple suspension was arguably a more effective street bike than most repli-racers. Weird looks or not, the R1100GS was a hit from the outset, and proved more popular than the R1100RS. Of the 80 A61 engines produced every day at Spandau in early 1994, 60 were destined for the R1100GS. There were only detail changes to the R1100GS over the next few years. The lower front mudguard was moved forward by 192mm (7.5in) for 1995, a three-way catalytic converter was standard from 1996, and there was a special BMW 75th Anniversary version for 1998. This year also saw the debut of the similar R850GS, with the A62 engine of the R850R. While the R1100GS was undoubtedly effective in its intended function, continual problems with the chassis cracking at the gearbox frame mount led to several interim modifications until the R1150GS appeared during 1999. After the introduction of the six-speed R1150GS, the R850GS continued unchanged to use up the supply of earlier components.

Below: From the beginning, the R1100GS was a success, providing outstanding all-round performance, particularly on the tarmac. *Australian Motorcycle News*

Right: Despite its size and weight, a strong rider could still manage to make the R1100GS work off-road. *Australian Motorcycle News*

THE R1100R & R850R

As the air-cooled boxer range limped into its final years of production, with minor variations on the existing theme, the range of new boxers was expanded. After 18 months, the new R259 boxer was proving outstandingly successful, and motorcycle production escalated during 1993 and 1994 to nearly 45,000, buoyed by the F650. The success of a new generation of naked bikes, such as Ducati's Monster and BMW's own R100R, prompted the release of the naked R1100R for 1995. There was also a similar, but smaller displacement version, the R850R, although this wasn't as popular.

Continuing a long BMW tradition of mix and match, the R1100R took the milder tuned, higher torque, engine of the R1100GS, placing it in the more sporting R1100RS chassis with R1100GS front and rear subframes. Without a fairing, this was termed a 'grass roots' motorcycle, and again, BMW had created a winner. The only difference between the powerplants of the R1100GS and R1100R was the location of the

The R1100R Telelever combined features of both the RS and GS to provide unique characteristics. The wire-spoked wheels and screen were options. *Australian Motorcycle News*

oil cooler, with the R1100R receiving two smaller oil coolers, mounted on either side above the cylinder head. The high-rise stainless steel exhaust system also came from the R1100GS.

New for the R850R was the 848cc A62 engine. While the lower end was the same as on the larger engine, there were 87.5mm pistons and smaller valves (32mm inlet and 27mm exhaust). The power was still a respectable 70bhp at 7,000rpm, and there was an electronically governed 34bhp version for graduated European driving licence regulations. Except for a lower final drive ratio on the R850R, both the R1100R and R850R shared the same running gear. By combining features of both the R1100RS and R1100GS, BMW engineers managed to create a unique motorcycle. The front and rear subframes were from

the R1100GS, as was the separately mounted handlebar, but this was a narrower, two-piece aluminium unit. As the R did not require such long travel suspension, the Telelever A-arm was from the RS, along with the front and rear shock absorbers. This combination of components resulted in slower steering geometry to both the R1100RS and R1100GS. The steering head angle was 27°, with 127mm of trail, while the wheelbase of 1,487mm was midway between the two parent models.

As there was no fairing to provide any front down force, a non-adjustable hydraulic steering damper was fitted between the slider bridge and the A-arm. Two types of wheel were also available for the R: the 17in and 18in cast-alloy wheels of the RS were standard and traditional cross-spoked wheels (18in front) were

With its upright riding position and high handlebars, the R1100R was a surprisingly sporting motorcycle. *Australian Motorcycle News*

available as an option. The front Brembo brakes were from the RS, while the rear brake came from the GS, and ABS II was an option with either set of wheels.

New styling set the R apart from its two parent models, the round headlight establishing a theme that was accentuated by the 21-litre steel fuel tank. The low and wide seat was adjustable for height, down to a low 760mm, so even the shortest-legged rider could feel comfortable. The standard equipment was relatively basic though, with no tachometer or clock, although these were available on the long list of options. The only real disadvantage of the R was that at 235kg (517lb) it was no lightweight. The lower-powered R850R was even more disadvantaged by the considerable weight, and proved so unpopular in some export markets that it was soon discontinued.

The rather hefty weight aside, the R1100R was an impressive motorcycle. If the R1100RS and R1100GS had not managed to convert the diehard traditionalist to the benefits of the modern boxer, the R1100R certainly did. Although disconcertingly wide to those used to older style narrow motorcycles, one ride on the high torque, easy-handling R1100R was enough to convince anyone that this was the best handling BMW boxer yet, and possibly the perfect all-round street motorcycle. Even the most cynical detractor had to admit that electronic fuel injection was not only here to stay, but was vastly superior to any carburettor. The styling was still unusual, but the R1100R set a new standard for naked motorcycles.

Changes were minimal until the R1100R was replaced by the R1150R for 2001. For 1996, there was

Opposite: The fourth variation
of the new boxer was the
R1100RT, and this proved an
immediate success. *Australian
Motorcycle News*

Below: The cockpit of the
R1100RT was perfect for
comfortable long-distance
sport touring. *Australian
Motorcycle News*

a standard catalytic converter, and the R1100R
received a redesigned cockpit for 1997 that included a
clock and a tachometer. There was a larger diameter
chrome-plated headlight, now with an aluminium
support. A special BMW 75th Anniversary edition was
available for 1998, and all R models again received a
moderate facelift. This year there was more chrome,
and several powder-coated silver components.
Throughout its six-year lifespan, the R remained very
popular, with nearly 54,000 R850R and R1100R
models produced. To use up the stock of R1100R and
R850R components after the introduction of the new
R1150R, a R1100R/R850R classic 'Special Model' was
produced. This ivory retro-style model featured wire-
spoked wheels and chrome-plated cylinder head covers.
While the R1100R Special Model ended production
during 2001, the R850R version continued for 2002.

THE R1100RT

The continual expansion of the new boxer line-up saw
the fourth incarnation, the R1100RT for 1996. Like
the R100RS-series, the R100RT, after a shaky start
back in 1978, garnered a strong following over its 18-
year production run. With its similar brother, the
R80RT, sales numbered over 57,000 by the time the
final R100RT Classic rolled out of the Spandau
factory. Yet, even though the old air-cooled models had
proved popular, the new versions totally eclipsed them.
Total sales of more than 53,000 for the R1100RT
between 1996 and 2001 almost matched the air-cooled
RT figures, and were vastly greater than the 28,000
odd for the R100RT alone.

There was a good reason why the new R1100RT
was so popular. Carrying on from the R1100R by

Opposite: Apart from the cruiser
styling, substantial chrome
plating and polished aluminium
distinguished the R1200C from
other boxers. *BMW*

incorporating components from the R1100RS and R1100GS, the R1100RT provided a level of touring comfort and seamless power that made it almost comparable to BMW's own K1100LT, rather than the sometimes lumpy R100RT. Using the 90bhp R1100RS engine, but with a lower final drive ratio to compensate for the larger fairing, the R1100RT, like the R1100R, used the front subframe of the R1100GS, with the Telelever A-arm, rear subframe, and shock absorbers of the R1100RS. As on the GS, the rear shock absorber also included the remote hydraulic spring preload adjuster. The wide two-piece handlebars were separately mounted, and isolated from the Telelever tilt as on the R1100GS and R1100R. As the R1100RT used a combination of suspension components, the steering geometry was similar to the R1100R, with a steering head angle of 27.2°, 122mm of trail, and a 1,485mm wheelbase.

The 17in and 18in wheels and 305mm front brakes were also shared with the R1100RS, while the rear disc was the smaller (276mm) unit of the R1100GS. Antilock brakes were standard, and the level of protection offered by the thermoplastic and rubber full fairing was extremely high. Emanating from the styling direction of David Robb, the small integrated rectangular headlight and large rounded panels may have taken some getting used to, but the plump fairing was aerodynamically excellent and worked well. A drag coefficient of 0.494 CdA with a flat windscreen and panniers was testament to this.

There was a new two-piece front mudguard designed to improve stability and reduce front wheel lift, while the windshield was electrically adjustable for rake by 22°, and height by 155mm (6in). Completing the touring specification was a large, 26-litre fuel tank, rather ineffective vents to supply warm air from the oil cooler to the rider's hands, and a three-way height-adjustable seat. As expected from BMW, a company with so much experience in producing touring and sport-touring motorcycles, the level of standard and optional equipment was comprehensive. Standard was a luggage rack, and 33-litre panniers, while the options included a cassette radio and a rear top-box.

Weighing in at 282kg (622lb), compared to the earlier R100RT, the R1100RT exuded an almost intimidating presence. Not only significantly heavier, the R1100RT was huge compared with the earlier incarnation. Yet, on the road, the R1100RT defied the

sceptics and was a remarkable long distance touring motorcycle, and one that was surprisingly agile away from the autobahn or highway. Apart from a 75th Anniversary special edition in 1998, there were few changes to the R1100RT before it was replaced by the R1150RT in 2001. For the 2000 Model Year there was a choice of two optional windshields, one higher, and one higher and wider, and as it was so popular it was a natural candidate for the updated 1,130cc engine and new Integral ABS during 2001.

THE R1200C & R850C

To many enthusiasts, one of the more unlikely renditions of the new boxer appeared in September 1997, the R1200C cruiser. However, as motorcycle industry observers were well aware, a cruiser was inevitable, as this was the fastest growing market segment in the early 1990s. In the world's largest motorcycle market, the USA, cruiser sales were outstripping sportsbikes three to one, and by 1997 numbered 150,000 units. To expand, BMW also needed a foothold in this lucrative area, and early in 1994 the new director of BMW Motorrad, Dr Walter Hasselkus, and the board of directors sanctioned the development of a cruiser.

The traditional cruiser was a high-torque, large-capacity, V-twin, and most other motorcycle manufacturers pursued this route. Not one for following others, BMW maintained their independent path, and it was quite a task to create an American-style cruiser out of the high-performance R1100RS and R1100RT. The first step was to enlarge the engine, with a 2mm bore increase to 101mm, and a longer stroke of 73mm. The displacement jumped to 1,170cc. Known as the A63, the compression ratio went down to 10:1, and as peak power was no longer a consideration, there were smaller valves (34mm and 29mm), and milder camshafts providing less valve lift. The throttle bodies were reduced by 15mm, to 35mm, and while the peak power was only 61bhp at 5,000rpm, the torque was 98Nm at a low, 3,000rpm. More significantly, the torque curve was flat in the important 2,500–4,500rpm range. Flywheel weight was increased, and the only real shortcoming of the larger engine was increased vibration. From the K1200RS came the smaller diameter (165mm) hydraulically actuated clutch, and the more compact transmission

housing. The gearbox was also based on the design of the K1200RS, but with only five speeds. Also from the K1200RS came the third-generation Motronic MA 2.4 engine management system which provided an automatic choke and diagnostic facility, and a catalytic converter was fitted as standard.

The design of the chassis was quite different to the four previous versions of the new boxer. Whereas these were all a product of mix-and-match, the cruiser style demanded something unique. BMW was already committed to the Telelever front suspension, and while the frame and A-arm could be comfortably hidden by large fuel tanks and fairings on the other four-valve boxers, on the cruiser these would need to be integrated into the styling. Thus, there was a new

aluminium front subframe, incorporating the dual oil coolers, and the longer longitudinal A-arm was of welded aluminium (instead of steel on the other boxers), and which was highly polished to accentuate the cruiser image.

The Telelever fork retained the usual, 35mm telescopic tubes, the long handlebars were separately mounted as on the GS, and the single front shock absorber provided 144mm of wheel travel. To suit the expectations of a cruiser the steering geometry was much more extreme than on the other boxers, with a steering head angle of 29.5°, with only 86mm of trail. The wheelbase was a gigantic 1,650mm (65in).

Some of the wheelbase length was obtained through the 90mm longer swingarm. Rather than the usual

Opposite: Like all BMWs, there was a wide range of options available for the R1200C, including screen and saddlebags. *Ian Falloon*

Above: The R1200C Independent for 2001 included a solo seat and cast aluminium wheels. *BMW*

subframes. Because there was no Paralever, rear wheel travel was limited to 100mm, and this, in combination with the longer swingarm, largely compensated for the negative effects of the shaft drive acting on the suspension. The original spring and rear shock absorber were very hard and from November 1998, a more comfortable unit was substituted.

Specific wire-spoked wheels were designed for the cruiser. Retaining the cross-spoke design that allowed the fitting of tubeless tyres, the front was a narrow, 2.50 x 18in and the rear a wider 4.00 x 15in wheel. The rims were chrome plated, while the front brakes came from the R1100GS and the rear brake from the R1100RS. As was now usual, ABS was also available as an option.

Charged with the cruiser's styling, David Robb went for a lot of chrome and polishing. From the A-arm to the rocker covers and air intake shrouds, the R1200C was certainly distinctive, and the quality of materials

Paralever set-up, this reverted to the earlier Monolever style, but with two universal joints like the Paralever, which was considered aesthetically difficult to incorporate in the cruiser image, but also, it did not allow enough room for the twin exhausts that were envisaged. Unlike earlier Monolevers, the single shock absorber was mounted centrally on the cast arm, behind the front pivot. The swingarm did not pivot on the gearbox housing (allowing the use of the more compact K1200RS unit), but on the tubular steel rear subframe. The engine was still a stressed member, supporting the pivots of the A-arm, and front and rear and execution was extremely high. There were new handlebar switches, moisture-resistant leather handlebar grips and seat, and a 17-litre steel fuel tank. The rather extreme cruiser riding position with its huge buckhorn handlebars was non-adjustable, but a flatter, drag-style bar was an option. For 1999, there was an optional 20mm thicker comfort seat. Not everyone took to the R1200C's extravagant styling, but its supporters far outweighed its detractors. During 1998, it was the best-selling BMW motorcycle, and this success continued. By the summer of 2001, production numbered more than 30,000. A similar R850C was offered during 1998, identical to the R1200C apart from the smaller displacement A62 engine. However, like the R850R, this proved unpopular and was discontinued for 2002. The cruiser buyer wasn't budget conscious, and there was no substitute for capacity with this type of motorcycle.

Avantgarde versions of the R1200C and R850C supplemented the Classic cruiser for 2000. In contrast to the heavily chromed appearance of the R1200C, the Avantgarde featured a black enamel finish for the engine and drivetrain, and graphitane (graphite and magnesium) for a host of previously chromed components ranging from the fork tubes, the A-arm and front and rear frames. Also setting the Avantgarde apart was a set of lower touring handlebars. In an effort to widen the appeal of the cruiser further, BMW followed the lead of other cruiser manufacturers by offering a third incarnation, the Independent (or Phoenix in the USA), for 2001. This had a solo seat, oval mirrors, new wheels, an additional headlight, and a small speedster-type handlebar fairing, to impart a 'bad boy' image. The aluminium wheels were two-piece, with three-spoke inner hubs replacing the usual cross-spoke wheels and were fastened together by titanium bolts.

There was the option of a high or lower handlebars, and even more chrome than on the Classic model. Chrome extended to the alternator cover, the air intakes for the oil coolers, and the brake and clutch reservoir covers. For 2003, the cast aluminium wheels were an option for the R1200C, and the Independent was available with a small passenger seat.

THE R1100S

Even as the sport-touring R1100RS was in the final stages of development, a more sporting boxer was envisaged. In 1993, a pure sports R1 engine with desmodromic valves was developed, but it was discontinued after a short test period. This 996cc boxer engine (98 x 66mm) featured a high-mounted crankshaft and two high water-cooled cylinders, with all the engine accessories underneath the engine. Each

cylinder featured twin overhead camshafts, driven by four chains, with four desmodromically actuated valves set in a very shallow combustion chamber. There were vertical intakes above the cylinders incorporating a Bosch injector and throttle position sensor, while the exhausts exited straight downwards. Installed in an aluminium twin-spar chassis with a Telelever front suspension system, the projected 140bhp boxer was not produced. Water-cooled cylinders seemed incongruous to the boxer concept as envisaged by Max Friz. He put the air-cooled cylinders out in the air stream for maximum cooling, and while this layout still provided perfect primary and secondary balance, there were the disadvantages of engine width, particularly in a modern sportsbike. As BMW remained committed to the boxer layout for traditional reasons it seemed more logical to base the sporting boxer on the existing A60 engine.

One of David Robb's first projects upon joining the BMW motorcycle division was to produce a sporting concept boxer prototype. This was displayed at the Cologne Show at the end of 1994, and spawned a rebirth of the 'S', not seen on a boxer since the demise of the R100S in 1979. The 'S' designation for BMWs was traditionally reserved for higher performing sports versions, and was initiated with the R50S and R69S of the 1960s, culminating in the classic R90S and R100S.

In the manner of the earlier classic R90S, the R1100S would be the most powerful incarnation of the boxer yet, with sharper handling and styling to match. Although not as radical as Robb's prototype, the 'S' was to be a refined sportster rather than an out-and-out street racer. BMW's designers were well aware that the large shaft-drive boxer could not match the powerful and light four-cylinder Japanese race replicas, but set about creating a more focused performance boxer that would excel as a high speed, long-distance machine. The R1100S was also representative of a new design philosophy which was evident with the K1200LT luxury tourer. This saw the creation of more extreme motorcycles, positioned outside the traditional BMW sport-touring mould.

Central to the R1100S was a hot-rodded 1,085cc A61 engine. Forged pistons with a 2mm higher dome provided an increase in the compression ratio, to 11.3:1, and allowed the redline to climb to 8,400rpm (from 7,900rpm). A reshaped inlet tract, less restrictive plate-type air filter, and a two-into-one-into-two stainless-steel exhaust system with integrated catalytic converter, saw the power rise to 98bhp at 7,500rpm. The exhaust system, which included much larger (45mm) headers, and two Ducati 916-style mufflers under the seat was said to be responsible for 70 per cent of the claimed power. Forged con-rods provided security at the higher revs, and the oil capacity was

Featuring the highest horsepower version of the boxer engine and top-quality chassis components, the R1100S emulated earlier sporting boxers. *Cycle World*

The R1100S twin muffler set-up was almost Ducati 916-like, but the Paralever was traditional BMW. *Cycle World*

DAKAR RALLIES AND THE F650GS

After a 12-year hiatus, BMW was back in the Paris–Granada–Dakar Rally with an official entry of four F650-based machines in 1998. There was little joy this year however, and only one machine finished the event (in 35th place), but BMW was back in 1999. Gottfried Michels and team manager Richard Schalber prepared four F650RR rally machines, with perimeter-style chrome-molybdenum frames and a titanium rear subframe. With twin Mikuni flat-slide carburettors the power of the 700cc single was boosted to a claimed 75bhp. The 28-year-old French rider, Richard Sainct, switched from KTM, and narrowly won the Granada–Dakar Rally. His BMW was heavier, but faster than the equivalent KTM, continuing a tradition in this rally established by BMW with the R80 G/S in the 1980s. Sainct was back on a BMW for the 2000 Paris–Dakar–Cairo Rally, and this year the BMWs totally dominated the event. Not only did Sainct win by a massive margin of 32 minutes, but F650-mounted Spaniard Oscar Gallardo finished second. Frenchman Jean Brucy was fourth, also on an F650, completing a BMW whitewash in this prestigious and gruelling event. This was the final year for works BMW singles, except for the entry for Andrea Mayer who finished 30th in 2001, the first woman home.

The success in the Paris–Dakar Rally coincided with the decision to move production of the F650 to Berlin-Spandau. A new assembly line was created, and for the 2000 Model Year, there was a revised F650GS, along with a special Dakar version. As before, all the components were outsourced, the only parts actually manufactured in Berlin were the camshafts, and these were then shipped to Bombardier-Rotax in Gunskirchen, Austria for installation in the engines. Along with the camshafts, BMW redesigned the cylinder head, incorporating features from the Motorsports M3 car, with the throttle butterfly very close to the injectors of the new BMW (BMS) electronic engine management system. There was a higher, 11.5:1 compression ratio, and with a large stainless steel dual muffler exhaust system, the power was increased slightly over the F650, to 50bhp at 6,500rpm. Along with the digital engine management set-up, the F650GS was also the first single-cylinder motorcycle to feature a standard catalytic converter.

Also new was the rectangular steel bridge-type frame with a bolted-on steel lower section and rear subframe. The 41mm Showa fork (with fork brace) and rear shock absorber were carried over from the F650, as were the 19in and 17in wheels, and Brembo brakes. A new type of ABS developed with Bosch was also available as an option. Weighing only 2.1kg (4.6lb), this more compact system was better suited for lighter motorcycles, and could be manually deactivated if required. Intended more for off-road use, the Dakar F650 came with a 21in front wheel, increased suspension travel (210mm front and rear), a higher (870mm) seat, and an F650RR windshield.

The styling of the F650GS was strongly influenced by the new R1150GS and F650RR. Underneath the re-styled plastic dummy tank hid the large air-box, and pressure-cast aluminium oil reservoir and coolant tank, with the 17.3-litre fuel tank located under the seat. Fuel filling was through an aircraft-type filler on the right. A wider range of options was also available for the F650GS, extending from a seat lowering kit, to heated handlebar grips, and an integrated luggage system. As it weighed a considerable 193kg (425.6lb), the F650GS was still more suited to the tarmac than the dirt, and even the 192kg (423lb) Dakar was a handful in difficult off-road situations. This did not detract from their popularity, with production of the F650GS and Dakar exceeding 30,000 units by the summer of 2001. The Dakar version accounted for more than 6,000 units and double shifts were required to meet demand. This year, the F650GS was the best-selling BMW motorcycle worldwide. For 2002, the Dakar was also available with optional ABS.

From the year 2000 the F650
was built in Berlin, and an
F650GS was created to celebrate
the F650 victories in the Dakar
rallies of 1999 and 2000.
Australian Motorcycle News

increased by 350cc to improve cooling. This was achieved through modified crankcases which soon found their way to other four-valve boxers. There was the new Bosch Motronic MA 2.4 injection system, a lighter (by 0.97kg) 600-watt alternator and 14Ah battery, while magnesium valve covers shaved a few additional grammes off the weight of the engine. This more powerful engine was then mated to the smaller clutch and six-speed gearbox of the K1200RS, and installed in a new four-piece frame. The front section included a lightened Telelever, the steel A-arm bolting to a die-cast aluminium frame. A welded aluminium main frame supported the swingarm pivot and top rear shock absorber mount, and there was a tubular steel rear subframe to support the seat and luggage rack.

For the 'S', the Telelever was developed further. The fork sliders were now machined rather than cast, imparting almost an upside-down fork look, but also contributing to the loss of about a kilogram of unsprung weight from the Telelever set-up. The front suspension unit was a single-sleeve gas-pressure damper, providing 110mm of spring travel. Rebound damping was adjustable by an innovative wheel between the fuel tank and upper fork bridge, enabling easy adjustment while on the move. There was the usual Paralever rear swingarm, the single shock absorber featuring a 40-position hydraulically adjustable spring preload. The new frame and suspension provided different steering geometry to the other five four-valve boxers, with a rake of 25°, 100mm of trail, and a wheelbase of 1,478mm (58in).

Another carryover from the K1200RS was the beautiful finely cast five-spoke wheels and front brakes with discs mounting directly on the wheel spokes. Not only were the wheels an aesthetic triumph, they were also technically superior to those on the R1100RS. A significant reduction in unsprung weight was achieved with the front wheel weight down to 5.59kg (from 6.05 kg) and rear wheel weight down from 6.47kg to 6.13 kg.

The front master cylinder diameter was reduced to 16mm (from 20mm on the R1100RS) to decrease braking effort, and the rear disc was the smaller, 276mm unit of the R1100R and R1100GS. There was even a carbon-fibre front mudguard, another first for BMW. A result of all this attention to weight saving saw the R1100S tipping the scales at 229kg (505lb), still no lightweight, but not too shabby when put into

the context of the earlier 220kg (485lb) R100S that was a lot less powerful, but which was still a formidable performer.

Designer David Robb intentionally set out to provide an environment where the rider felt part of the motorcycle, and not perched on top as with many sports bikes. An 18-litre, almost sculpted, aluminium fuel tank lay underneath the flowing plastic bodywork that virtually enveloped the rider. The four-piece fairing was developed in the wind tunnel and incorporated turn signals and hand protectors, and featured an attractive, and distinctive automotive-type of ellipsoidal dual headlight incorporating an asymmetrical large, low beam (H7) and a smaller high beam (H1). The clip-on handlebars bolted directly to the vertical tube beneath the fork bridge and accentuated a sport-touring riding position, rather than aggressive sport. On the road, the R1100S was undoubtedly the sharpest steering and most sure-footed boxer yet.

Once underway there was no evidence of the weight and here was a boxer that could use all of its 50° lean angle (up from 49° on the R1100RS). If required, there were also options, including higher-comfort handlebars and a taller windscreen, ABS, a wider, 5.50 x 17in rear wheel, steering damper, true rear-set footpegs, black engine covers, and a chin spoiler. For those requiring a sportier orientation, the sports package comprised a handlebar damper and longer shock absorbers which raised the motorcycle by 18mm at the front and 20mm at the rear. The maximum angle of lean increased to 52°, and just to show how efficient they were, BMW included a longer side stand in the kit. And as always, there was the option of BMW's hard luggage, which was still the best available. The R1100S was surely the finest sport-touring motorcycle obtainable, especially if the required emphasis was on sport rather than touring. Many buyers thought so, and production had amounted to more than 20,000 by Summer 2001.

THE R1150GS

The R1100GS received what was termed 'mid-life freshening' during 1999. Despite a tarnished durability image, sales of the four-valve GS models over their six-year lifespan numbered over 45,000. It made sense to incorporate some of the developments of both the R1200C and R1100S in a replacement version. To provide an even flatter torque curve, the engine

Opposite: Carrying on from
the successful R1100GS was
the considerably revised
R1150GS for 1999. *Australian
Motorcycle News*

displacement was increased to 1,130cc with the 101mm cylinders of the R1200C cruiser (without polished fins), but with higher compression (10.3:1) pistons. From the R1100S came the cylinder heads, lighter magnesium valve covers, and crankshaft, but there was a new camshaft designed to further optimise the torque curve. Like the R1100S, a larger diameter exhaust system (45mm) was responsible for much of the increase in power to 85bhp at 6,740rpm. The exhaust system was also restyled to emulate older boxers, the twin chrome-plated stainless-steel header pipes arcing under the cylinders. They were then joined by a cross pipe, a double tube feeding into the pre-muffler, and a redesigned muffler. Other features shared with the newer models were the Motronic MA 2.4 engine management system, and a standard, fully controlled three-way catalytic converter. Cooling was also uprated through the larger oil cooler of the R1100RT, and like the R1100S, a lighter, 600-watt alternator and small, 14Ah battery were standard (although the 700-watt and 19Ah battery were standard if heated handlebars were specified).

Also from the R1100S came the smaller diameter (165mm) hydraulic clutch and six-speed transmission. Although the lower five ratios were identical, the R1150GS featured a higher sixth gear, designed as an overdrive for more comfortable touring. The six-speed gearbox was certainly an improvement over the sometimes-recalcitrant five-speed unit, but it still wasn't perfect and would be replaced by a revised transmission for 2003.

A significant improvement over the R1100GS was the revised three-piece frame set-up, a development of the existing R1100RS and R1100GS type, rather than the R1100S's twin-spar top frame. Although the engine and transmission housing were still load-bearing components connected directly to the front and rear subframes, the rear set-up was revised considerably. The longer gearbox housing was strengthened, and an outboard aluminium casting, which supported the rear subframe, reinforced the swingarm pivot. Not only did these two outboard plates provide additional rigidity and footpeg supports, but they were designed to finally overcome the problems of rear frame failure in arduous conditions. Because the six-speed transmission case was longer than that of the previous five-speed unit, the Paralever swingarm was shortened from 520mm to 506mm to maintain the same wheelbase as before.

The front Telelever was also modified, incorporating the lighter-assembled A-arm of the R1100S rather than the previous cast type, and machined fork sliders. These also imparted an upside-down fork look, and while this was only an illusion, it did provide the R1150GS with a more conventional front-end appearance. However, both the A-arm and single shock absorber were still strongly evident, with both the front and rear shock absorbers carried over from the previous R1100GS. Also shared with the earlier model were the cross-spoked wheels (with a modified hub at the rear), and Brembo brakes. The front brake pads were now sintered metal rather than organic.

David Robb set out to create a distinctly different machine from the R1100GS, and while the dimensions were similar, the two designs could not be mistaken. While retaining two front mudguards, the lower front guard was now inconspicuous black, matched by a redesigned back wheel cover. Although still beak-like, the front upper mudguard served as an oil-cooler air scoop and wind deflector. It was also now integrated into the 22.1-litre fuel tank. Quite different to the previous rectangular headlight were the two asymmetrical ellipsoidal headlights, similar to those of the R1100S. These also featured a large H7 bulb for low beam and a smaller, free-form H1 bulb for high beam. Completing the new look was a three-position windshield that could be removed if required. The separated handlebars incorporated the new generation switches, and vibration-isolating rubber handgrips, and were even wider than before at 903mm. These provided incredible leverage, belying the considerable 249kg (549lb) weight. Although the seat could be adjusted to two positions (840mm or 860mm), the R1150GS was still an intimidating machine for a shorter rider. This was even more pronounced if taken off road, where the GS was a remarkably good performer but it took a strong person to control if it fell over. Less of a problem on the tarmac, and just like the earlier GSs, the R1150GS could surprise many a sportsbike on a twisty road. The R1150GS continued the success of earlier GSs, with more than 32,000 produced by the summer of 2001. For 2003, the R1150GS received the updated six-speed gearbox of the R1200CL, and optional, partial Integral ABS that could be deactivated if required.

Coinciding with the release of the R1150GS, two works GS/RRs were prepared to run alongside the

F650RRs in the 2000 Paris–Dakar–Cairo Rally. As with the rally-winning GSs of the 1980s, these were prepared by HPN. One displaced 900cc, and weighed 190kg, while the other was a full 1,100cc. Running on carburettors rather than fuel injection (due to the dubious fuel in Africa), and with enough power to top nearly 200km/h (125mph), conventional White Power upside-down telescopic forks were fitted at the front instead of Telelevers. At the rear was a special Paralever swingarm with a PDS shock. British rider John Deacon rode the '1100', crashing out, while *Cycle World* staffer, Jimmy Lewis, took the R900RR into third position in the gruelling rally, his boxer running the entire race without an engine strip down.

Lewis was back on the works R900RR for the UAE Desert Challenge, the Dubai Rally, which is a prelude to the 2001 Paris–Dakar Rally. Although largely unchanged, a more aerodynamic fairing, lower fuel tanks for improved weight distribution, and new suspension made the R900RR a rally winner. Lewis provided the boxer with its first race victory since the days of Gaston Rahier and Hubert Auriol. Expecting to repeat this in the Paris–Dakar Rally, the event turned out to be slower and more tortuous than expected.

Created as the ultimate city transport, the C1, with its integral roll cage, was designed to provide motorcycle manoeuvrability with automotive safety. In Germany it could be operated without a motorcycle helmet. *BMW*

Opposite top: Almost everything was new on the R1150R, while the chassis was similar to that of the R1150GS. *BMW*

Opposite bottom: The R1150R was a successful development of the R1100R, with more purposeful styling. *BMW*

THE C1

A design study initiated in 1989 called for a vehicle suitable to overcome the severe traffic congestion in the central zones of large cities. This led to BMW producing the prototype C1, providing safety comparable to a car, but with motorcycle mobility. A joint development by BMW Technik and BMW Motorrad, the C1 design study was first presented at the Cologne Show at the end of 1992. With its integral roll cage, safety-cell, and seatbelt, BMW envisaged the C1 to be ridden without a helmet and on a car driving licence. The C1 remained a fantasy until the end of 1997 when BMW announced it would implement development as the German authorities had agreed to its use without a helmet. As a further incentive, drivers who obtained a car licence in Germany before April 1980 could operate the C1 without a motorcycle licence, and there were exemptions from road tax.

The C1 finally materialised for the 2000 Model Year, with a liquid-cooled double-overhead camshaft four-valve single-cylinder four-stroke engine producing 15bhp. For 2001, there was also the 176cc C1 200, producing 18bhp. Compared with many city-type vehicles, the C1 boasted a very high specification. There was Telelever front suspension, Brembo disc brakes front and rear, and optional Bosch ABS. The C1 was also very manoeuvrable with its small, 13in and 12in wheels, and was easy to ride having a centrifugal clutch, continuously variable transmission and belt final drive. The weight was 185kg (408lb), and the C1 200 was capable of around 120km/h (75mph). The engine could not be started without the rider strapped into the safety cell, but in the UK, most of the C1's appeal was negated through the requirement to wear a crash helmet.

Spaniard Juan Roma was well placed until he crashed, and John Deacon managed sixth place, winning the twins class. Jimmy Lewis crashed, breaking a collarbone, but somehow managed to finish the rally in seventh.

THE R1150R & R850R

In response to the backlash against repli-racers, nicknamed 'yoghurt-cups' in Germany because of their extreme multi-coloured plastic bodywork, David Robb restyled the naked R1150R to emphasise its elemental nature. The 1,130cc 85bhp engine came directly from the R1150GS, including the Motronic MA 2.4 electronic engine management and magnesium valve covers with R1100S cylinder heads and crankshaft being used. Also from the R1150GS came the six-speed gearbox, with the lower sixth gear ratio of the R1100S. The stainless-steel exhaust, with catalytic converter, was also similar to the R1150GS. Compared with the previous R1100R, the final-drive ratio was higher, to provide more relaxed high speed cruising, and the twin oil-coolers were now integrated into ducts at the side of the fuel tank.

The chassis of the R1150R was continued from the R1100R in that it was also an amalgam of components from the R1100RS and R1150GS. The basic frame layout was that of the R1150GS, with the Paralever swingarm supported by two additional aluminium plates that also served as footpeg mounts. To retain the same wheelbase and steering geometry as the successful R1100R, the front steel A-arm was based on the street R1100RS, but built out of steel tubing to enhance aesthetics. The Telelever fork tubes were not machined to imitate conventional upside-down forks and the Paralever swingarm was shortened by 14mm to compensate for the longer transmission housing. Although the spring travel on both the front and rear shock absorbers remained unchanged, the units were a new type. The front shock incorporated a single sleeve damper, and through an adjustment bolt on the A-arm, infinite adjustment of rebound damping was available. The rear shock absorber also provided hydraulic spring preload adjustment, while both shocks featured linear compression damping (rather than progressive as on the R1100R). Reflecting the trend towards wider, more sporting tyres, only cast, 17in aluminium wheels were offered on the R1150R. These were the five-spoke type from the R1100S, the rear wheel growing to 5.00 x 17in, allowing for a 170/60ZR17 rear tyre.

Complementing the new wheels were larger EVO front brakes. With the disc rotor size up to 320mm, and upgraded Brembo-Tokico four-piston calipers, not only was brake power increased, but the lever pressure was claimed to be 15 per cent less than before. Unsprung weight was reduced by 10 per cent, and the new, sintered brake pad life was extended by 50 per cent. It was also possible to change the brake pads in situ, without dismantling the caliper as was required previously. The rear brake was identical to that of the R1100R, and ABS was available as an option. This was the new Integral ABS system as first specified on the K1200LT, but was a 'partial' variant with the foot pedal operating only the rear caliper, while the linked front lever worked all three discs. This partially integrated system was designed for more sporting applications The reduction in unsprung weight from the wheels and brakes negated the necessity for the R1100R's steering damper.

Compared with the R1100R, the styling of the R1150R was fresher, and quite different. Apart from the round headlight and the instrument layout, everything else was new. This included the radical front mudguard (although the rear part came from the R1100RS), and the larger, 20.4-litre fuel tank. The seat height was 800mm, but a lower, 770mm seat was an option. The entire riding position was lower, further enhanced by the wider and lower handlebar. The only impediment to unsurpassed sportsbike performance was the weight of 238kg (525lb). However, the

Opposite: With a more powerful engine, and updated wheels and brakes, the R1150RT was an even more accomplished touring motorcycle than its predecessor. *BMW*

Below: The R1150RS looked very similar to the first R1100RS of nearly a decade earlier, but benefited from developments to the engine, transmission, wheels, and brakes. *BMW*

R1150R continued to provide excellent all-round performance, and remained one of the most appealing naked bikes available.

There were only minimal developments made over the next few years. The new, six-speed gearbox was fitted from 2003, along with braided steel brake lines. The R850R also made a return this year; identical to the R1150R but for the 70bhp or 34bhp 848cc engine. The six-speed transmission featured a lower sixth gear.

The R1150R also spawned a concept Roadster Boxer, designed by David Robb's Motorcycle Design Team, which was displayed at the Munich Intermot towards the end of 2002. Reflecting a trend towards more sporting and aggressive naked bikes and aimed at the younger rider, this incorporated some components

from the R1100S, like the Telelever fork legs and wider rear wheel, plus R1150GS headlights. As the public response to the concept roadster was so positive, in October 2002 BMW announced this model would enter production during 2003. Now titled the Rockster, the new model incorporated the new dual spark plug cylinder head that would feature on all 2004 boxers. There was also a special 80th Anniversary edition Rockster, with 2003 numbered examples produced.

THE R1150RT

Alongside the R1150R for 2001 was an updated R1150RT. As the R1100RT was one of the most popular models in the range, it was logical to

The single-sided swingarm with belt final drive was a unique feature of the F650CS. *BMW*

THE F650CS

Continuing a tradition which proved they were not afraid to redefine existing concepts, BMW released the astounding F650CS (City Sport) in 2002, replacing the mundane F650ST. Unlike the F650ST that was very similar to the F650 Funduro, the F650CS was a distinctive creation. Intended primarily for city riding, the F650CS also incorporated unique solutions to the problems of storage on this type of motorcycle. The BMW-developed Rotax engine and five-speed gearbox were from the F650GS, but because of air-box and exhaust modifications, the power of 50bhp was at a higher 6,800rpm, with a slight increase in torque to 62Nm at 5,500rpm (up from 60Nm at 6,000rpm). There was a new exhaust system, with the catalytic converter and oxygen sensor integrated into the silencer. There was no need for a pre-silencer which allowed the designers more freedom in styling the muffler so as to coordinate it better with the rest of the motorcycle.

There was also a larger cross-section rectangular steel frame, which reverted to include the engine oil reservoir, as on the first

F650. The two upper arms of the bridge-type frame were connected to one another at their lower end, holding 2.5 litres of oil. This was done to provide an integrated storage compartment in the central fairing and dummy tank above the air-box, as the 15-litre fuel tank remained under the seat as on the F650GS. Featuring a translucent fastening rail on both sides, this compartment could house a special waterproof, 12-litre soft bag, plastic hard case, helmet, or audio system with two integrated, weatherproof loudspeakers.

Another new feature was the aluminium single-sided swingarm with toothed-belt final drive, continuing a path of innovative wheel guidance systems. The toothed-belt-drive was particularly suitable for street motorcycles, as it required no lubrication, and boasted superior durability. The toothed drive belt was 26mm wide, with an 11mm partition. Like the F650GS, the F650CS had a central rear shock absorber, with a rising rate linkage, but with less (120mm) spring travel. To provide better

Accessory soft bags, and the provision for storage in the dummy fuel tank, provided the F650CS with a new dimension in practicality. *BMW*

street operation, the Showa fork also provided shorter spring travel of 125mm. Completing the chassis components were new, 17in three-spoke cast-aluminium wheels with a 'curved' design from the hub to the rim to accommodate the brake and toothed-belt sprocket. From the F650GS came the 300mm single front disc with twin-piston floating caliper, while the rear brake was a single, 240mm with a single-piston floating caliper. ABS was optional, and was selected by more than half of all purchasers.

Completing this radical design concept was styling that included a greyish-blue tinted, clear-view windshield, a translucent windshield support, and dual ellipsoid headlights. A standard rear

luggage rack incorporated aluminium grab handles and accommodated an accessory soft bag that could connect to another over the passenger seat, creating 62 litres of storage. And if all that wasn't enough, the range of options allowed customers to choose from three base colours, two fairing inset colours, and two seat colours, for ultimate personalisation. Further options extended to a lower (750mm) seat, GRP engine spoiler, and an on-board computer.

Although low and narrow, making it a perfect city motorcycle, out of the city environment the F650CS was limited by its moderate power and 189kg (417lb) weight. Stability was ensured through the long, 1,493mm (58.7in) wheelbase, but there was no denying the practicality of the well-thought-out luggage and low-maintenance belt drive. The riding position, accentuated by a low seat and wide handlebars, emulated that of a dirtbike in that the rider sat 'in' the bike rather than 'on' it. But in the city, where it was intended to be used, the F650CS provided all the advantages of both a motorcycle and a scooter. Scooter manoeuvrability, with motorcycle performance.

Opposite: The R1150GS
Adventure provided the
ultimate platform for any
serious motorcycling in remote
areas. *BMW*

incorporate updates to this model ahead of the long-lived, but less popular R1100RS. For the R1150RT the engine was also increased in size, to 1,130cc, using the R1150GS cylinders, but with higher compression, 11.3:1 pistons. Also using the Motronic MA 2.4 engine management system, the power increased to 95bhp at 7,250rpm, with torque up to 100Nm at 5,500rpm. More than 90Nm was available between 3,000rpm and 6,500rpm, providing a wide powerband. The power was transmitted through the smaller hydraulic clutch and six-speed gearbox of the other '1150' models, with sixth gear an overdrive, like on the R1150GS.

The chassis and suspension geometry were similar to the R1100RT. The steering head angle was an almost identical 27.1°, with the same 122mm of trail. The shock absorbers were also unchanged, providing identical travel, and the rear spring load featured an hydraulic adjuster. The longer, six-speed gearbox housing required a modified Paralever support, and a reduction in swingarm length from 520mm to 506mm to maintain the 1,485mm wheelbase. As with the R1150R, there were new supports for the rear shock absorber and redesigned footrest plates. One of the most significant developments was to incorporate the lighter and more modern-looking 17in wheels of the R1100S and R1150R. While the front wheel dimensions were unchanged, the rear wheel was a much more suitable 5.00 x 17in, allowing for the use of the latest generation of tyres.

On the front wheel was the new EVO brake, with the rear wheel featuring the same brake as on the R1100RT. Standard on the R1150RT was the Integral ABS braking system, but unlike the R1150R, this was the fully integrated version as fitted to the K1200LT, and more suitable for a tourer: the handbrake and footbrake levers both acted simultaneously on the front and rear wheel brakes. There was no longer any pulsing through the brake lever as on the earlier ABS II, but the system came under some criticism as it was tailored more for inexperienced riders rather than the seasoned traditional BMW pilot.

Completing the upgrades was a facelift, the new styling effectively silencing critics of the rotund R1100RT. Two fog lamps were integrated at the sides of the two tandem headlights, with the headlight adjustable for height by a hand wheel in the cockpit. The upper part of the fairing was redesigned, along with the front mudguard, but the windshield had the

same dimensions as before. This was also adjustable electrically through 22°, and 155mm via a switch on the left handlebar controls. The fairing included a warm airflow duct, air flowing from behind the oil-cooler to the cockpit. Holding 25.2 litres, the plastic fuel tank provided an ample cruising range. Other improvements extended to the seat, the two-piece, three-way height-adjustable seat offering improved comfort, particularly for the rider. The result of these developments widened the appeal of the RT even further. Offering almost as many luxury features and options as the full touring K1200LT, the R1150RT provided almost sport-touring capability when it was compared to its 100kg (220lb) heavier brother. Here was a touring motorcycle that worked perfectly anywhere, no matter what the road or the weather, and 11,000 buyers thought the R1150RT was the perfect touring motorcycle in its first full year of production.

THE R1150RS

Although the least popular model in the new boxer line-up, the R1100RS always maintained a loyal following from those interested in carving miles and apexes. Less bulky than the RT, but not as extreme as the 'S', the 'RS' still filled a niche so it was inevitable that it too would share in the developments of the other 1,100cc boxers. For 2002, the R1150RS replaced the long-serving R1100RS. The 95bhp engine was identical to that of the R1150RT, as was the six-speed gearbox with overdrive sixth gear, and the catalytic converter. The clutch used was also the smaller hydraulic unit.

The chassis and steering geometry were unchanged, with identical shock absorber travel, although the Telelever telescopic tubes were from the R1150RT. The rear shock absorber and shorter swingarm were also from the R1150RT, but the rear swingarm pivot points were altered to suit the different characteristics required for the 'RS'. The footrests came from the R1150R. The 17in five-spoke wheels were new, now being shared with all the street models except the cruiser. The front EVO brake was also a new design. Unlike the R1150RT, the Integral ABS was an option, and it was the more sporting orientated partial set-up that was also featured on the R1150R.

Visually the R1150RS was akin to the R1100RS. The fairing and rectangular headlight were similar, but the windshield was 80mm higher and 60mm wider,

increasing the area by 30 per cent. As in the past, the windshield was easily adjustable for inclination. The previously optional full fairing, which continued underneath the cylinders and enclosing the engine, was now standard. Other new features were the handlebar controls, and the front white direction indicator lights, while the wide range of options extending from luggage to chrome-plated cylinder protection covers remained as before. Carried over from the R1100RS were the rubber-mounted handlebars which still conveyed a feeling of imprecision. Considering that all the other new boxers had separately mounted handlebars, this seemed incongruous, and detracted from its ultimate sporting ability. Another feature of the R1150RS was the 285mm rear disc brake, as used on the R1100RS, while most other R-series models used a 276mm disc. Compared to the updates on the 'GS', 'R', and 'RT', the R1150RS was less intensive, and possibly an interim design. Even before its release, rumours abounded of a new motor, and a further restyling.

THE R1150GS ADVENTURE

Expanding the GS line-up was the R1150GS Adventure for 2002. Designed as the ultimate go-anywhere motorcycle, this was either the perfect desert motorcycle, or one for the ultimate Walter Mitty outback dreamer. Whatever the intended use, with its optional huge, 30-litre fuel tank, and 105-litre aluminium baggage system, the Adventure raised the ante for the size of off-road motorcycles. If the R1150GS seemed intimidating to smaller riders, the 253kg (558lb) Adventure was even more so.

The 1,130cc engine was shared with the R1150GS, but as the Adventure was designed for use in any part of the world, an optional coding plug for the Motronic MA 2.4 activated an alterative ignition map to allow the engine to run on regular, 91-octane fuel. The sixth-gear ratio was also shortened (as on the R1100S), and there was an optional lower first gear to improve

THE BOXER CUP AND
BOXER CUP REPLICA

Following a tradition of celebrity boxer support races for the Daytona 200 during the mid-1990s, the French and Belgian BMW distributors inaugurated the Boxer Cup in 1999. As support races for the Motorcycle World Championship, by 2001, the Boxer Cup series for R1100S motorcycles grew to include seven events. Few modifications were allowed, and the machines were essentially stock, but for racing exhaust systems and the sports package of a wider rear wheel and longer suspension. The enormous prize money (225,000 Euros), attracted many former professional racers, including Randy Mamola, Kevin Schwantz and Luca Cadalora, as did the chance of winning a BMW M Coupé, or R1150GS Adventure. The third prize of a C1 may not have been quite as actively contested.

From a field of 32 riders, former Belgian World Superbike star Stéphane Mertens won the series in 2001. Expanded to eight races for 2002, and receiving FIM and UEM cup status, tight racing saw seven different winners in the first seven races. Ultimately, Mertens took out his second International Boxer Cup, ahead of Thomas Hinterreiter and Andy Hoffman. For 2003, there were nine rounds, the first at the Daytona 200, followed by one at the British Superbike Championship round at Oulton Park, five with MotoGP, and two with the Masters of Endurance and World Endurance Championship.

The success of the International Boxer Cup series led to the release of the R1100S Boxer Cup Replica for 2003. Featuring paintwork modelled on that of the racing machines, complete

Opposite: Although still essentially an R1100S, the Boxer Cup Replica was lighter and featured more sporting suspension. *BMW*

Above; Setting the Boxer Cup Replica apart from the R1100S Boxer Cup was the Randy

Mamola signature and carbon-fibre reinforced valve covers. *Ian Falloon*

with Randy Mamola signature, the engine and the drivetrain was standard R1100S, but with a few performance enhancing modifications. The Replica included the sports suspension package and wider rear wheel, along with carbon-fibre reinforced plastic valve covers, engine spoiler, and seat cowling. Shared with the R1100S were the larger front brake discs, EVO brake calipers, and optional partial Integrated ABS. In addition to the Boxer Cup Replica, there was a plainer Boxer Cup, without the special decals, spoiler, seat cowl or carbon valve covers. This retained the sports suspension and wider wheel. Neither the Boxer Cup, nor Replica, featured a centrestand. Even though the claimed weight for both Boxer Cup editions was the same 229kg (505lb) as the R1100S, the Boxer Cup was undoubtedly lighter, and the sharpest handling BMW yet.

manoeuvrability in difficult terrain.

There were a number of developments to the chassis to provide increased suitability in tough, remote terrain. The front spring travel was increased 20mm to 210mm, while a new White Power rear shock absorber provided 220mm of travel. The damping also increased proportionally to spring deflection. There were no changes to the cross-spoked wheels, but the front brakes now included the EVO calipers, but with braided steel brake lines. The brake disc diameter remained at the earlier 305mm, and the optional ABS was the older ABS II system that could be switched off for off-road use.

A special seat, designed for hours in the saddle, was new for the Adventure. The rider could either sit forward for slow, rough going, or back for extended touring. There was a larger windshield, longer and wider front mudguard, handlebar protectors, and a huge aluminium bash plate under the engine. Options extended to engine protection bars, steel protection covers for the headlights, and knobby off-road tyres. Introduced as a mid-2002 model, for 2003 the Adventure incorporated a few developments. The six-speed transmission was updated, and there was now a Showa rear shock absorber and the optional ABS was the partial Integral ABS as featured on the more sporting boxers.

THE R1200CL

Although it appeared that every sector of the touring market was well covered with the both the R1150RT and K1200LT, 2003 saw the release of a further variant, the R1200CL (Cruiser Luxury). Based on the R1200C cruiser, this was intended as an American-style cruiser that could also swallow up miles in comfort, steering a different path to its touring brethren.

There was more to the R1200CL though, than simply bolting on touring equipment to the existing R1200C. The 'CL' was the first boxer to receive the new six-speed transmission which soon found its way to the entire boxer line-up, except on the five-speed R1200C. The previous six-speed gearbox was never well liked, many actually preferring the earlier five-speed version, and the new gearbox was an attempt to silence the critics. While the previous basic structure was retained, the geometry of the new gears was modified, extending further upwards and providing a larger overlap. This resulted in a softer gear mesh, with less running noise. There were also developments to the gearshift operation through modification of the shifting forks and gear recesses. Not only did the gearbox mesh more precisely, but there was a reduction in weight of around 1kg (2.2lb).

Apart from the R1200C engine, rear-wheel drive, cast-aluminium front frame, and tiny, 17.5-litre fuel

Left: The handlebar-mounted fairing, and four headlights of the R1200CL imparted a new image. *Ian Falloon*

Opposite: While the R1200CL was based on the R1200C cruiser, the integrated luggage and wide, 16in front wheel signified a different focus. *BMW*

tank, most of the R1200CL design was new. Even when compared with the R1200C the steering geometry was extreme, with the front wheel raked out an incredible 33.5°, with an equally astonishing 184mm of trail. The wheelbase of 1,641mm (64.6in) was comparable to the cruiser. There were new, flatter, Telelever and wider forks legs to provide this cruiser geometry and accommodate the wide, 150/80 x 16in front tyre. The 144mm front shock absorber travel was the same as the cruiser, while the similar Monolever rear suspension set-up provided 20mm increased travel to 120mm. The entire rear frame and swingarm were new, while the Monolever was reinforced and modified to accept the K1200LT rear disc brake. The new rear

frame was strengthened to accept footboards and luggage supports, and there were newly designed wheels, a 15in rear complementing the 16in front. The front brakes were the smaller, 305mm discs, with EVO calipers, and fully integrated ABS, as on the R1150RT, was an option.

Undoubtedly the most distinctive aspect of the R1200CL was the handlebar-mounted fairing which incorporated four headlights, two larger H4s for low beam and two smaller H1s for high beam. The alternator was uprated to 840 watts to power the extra lights, with a low-maintenance 19Ah battery. Although non-adjustable, the windshield was designed with a centre dip to provide a clear view of the road in all conditions, with the upper edge contoured to deflect air away from the rider. Two additional fairings on the side of the fuel tank were designed to create the look of a totally integrated fairing, and there was a deeply valanced front mudguard. Completing the touring cruiser image were wide handlebars, a stepped seat, and colour-coordinated integrated hard luggage. As expected of a luxury cruiser, there was a wide range of options, ranging from cruise control and heated handlebars, to intercom and CB radio. Undoubtedly, the R1200CL provided enough luxury to warrant its claim, along with the usual BMW ability to cover long distances in comfort (small fuel tank not withstanding). Yet, with a substantial weight of 308kg (679lb), and considerably more when fully loaded, the 61bhp engine was surely to be taxed to the limit, even if it did possess a wide, flat torque curve. The R1200CL also spawned a cruiser during 2003, the Montauk. Designed to fill a gap between the R1200 Classic and R1200CL, this featured an R1200CL chassis with minimal cruiser equipment. The engine had twin-spark cylinder heads, and additionally there was a new five-speed gearbox.

The first 80 years of BMW motorcycle production has seen the company rise and fall through several dramatic periods. The company almost died after the Second World War, and in 1959, and motorcycle production gradually slumped through the 1960s. Throughout this often-turbulent history, it has been the adherence to the traditional boxer engine layout that has seen the BMW motorcycle survive. With an increased proliferation of models, and the continual expansion of the range to meet new niches in the marketplace, production levels are now higher than ever. Around 75,000 motorcycles have rolled out of the Spandau factory every year since 2000, a massive increase considering less than 5,000 emanated from the final year in Munich, a little over 30 years earlier. Unlike the days gone by, BMW motorcycles are no longer outside the mainstream. In recent years, BMW has shown it can adapt to the modern competitive, and often fickle, motorcycle marketplace, yet still offer a unique product. The future will undoubtedly see new concepts, along with technical innovation, to further widen their appeal.

APPENDICES
APPENDIX 1
BMW MOTORCYCLE PRODUCTION 1923–2001

Model	Year	Number	Model	Year	Number
R32	1923-26	3,090	R35	1937-40	15,386
R37	1925-26	152	R20	1937-38	5,000
R39	1925-27	855	R23	1938-40	8,021
R42	1926-28	6,502	R51	1938-40	3,775
R47	1927-28	1,720	R66	1938-41	1,669
R52	1928-29	4,377	R61	1938-41	3,747
R62	1928-29	4,355	R71	1938-41	3,458
R57	1928-30	1,005	R75	1941-44	18,000 approx
R63	1928-29	794	R24	1948-50	12,020
R11	1929-34	7,500	R25	1950-51	23,400
R16	1929-34	1,006	R25/2	1951-53	38,651
R2 Series 1	1931	4,161	R25/3	1953-56	47,700
R2 Series 2a	1932	1,850	R51/2	1950-51	5,000
R2 Series 2/33	1933	2,000 approx	R51/3	1951-54	18,420
R2 Series 3	1934	2,077	R67	1951	1,470
R2 Series 4	1935	2,700	R67/2	1952-54	4,235
R2 Series 5	1936	2,500	R67/3	1955-56	700
R4 Series 1	1932	1,101	R68	1952-54	1,452
R4 Series 2	1933	1,737	R50	1955-60	13,510
R4 Series 3	1934	3,671	R69	1955-60	2,956
R4 Series 4	1935	3,651	R60	1956-60	3,530
R4 Series 5	1936-37	5,033	R26	1956-60	30,236
R3	1936	740	R50/2 (incl. R50US)	1960-69	19,036
R12	1935-42	36,000	R60/2 (incl. R60US)	1960-69	17,306
R17	1935-37	434	R27	1960-66	15,364
R5	1936-37	2,652	R50S	1960-62	1,634
R6	1937	1,850	R69S (incl. R69US)	1960-69	11,317

Model	Year	Number	Model	Year	Number
R50/5	1969-73	7,865	R80RT	1984-95	22,069
R60/5	1969-73	22,721	K75C	1985-90	9,566
R75/5	1969-73	38,370	K75S	1985-95	18,649
R90S	1973-76	17,465	R65	1985-93	8,260
R90/6	1973-76	21,097	K75	1986-96	18,485
R75/6	1973-76	17,587	K100LT	1986-91	14,899
R60/6	1973-76	13,511	R100RS	1986-92	6,081
R100RS	1976-84	33,648	R100GS	1987-96	34,007
R100S	1976-78	11,762	R80GS	1987-96	11,375
R100/7	1976-78	12,056	R65GS	1987-92	1,727
R75/7	1976-77	6,264	R100RT	1989-96	9,738
R60/7	1976-78	11,163	K1	1989-93	6,921
R80/7	1977-84	18,522	K100RS (4-valve)	1989-92	12,666
R100RT	1978-84	18,015	K75RT	1989-96	21,264
R100T	1978-80	5,463	R100R and Mystic	1991-96	20,589
R45	1978-85	28,158	R80R	1992-94	3,593
R65	1978-85	29,454	K1100LT	1991-98	22,757
R100	1980-84	10,111	K1100RS	1992-96	12,179
R100CS	1980-84	4,038	R1100RS	1993-2001	26,037
R80 G/S	1980-87	21,864	F650	1993-2000	50,990
R65LS	1981-85	6,389	R1100GS, R850GS	1993-2000	45,870
R80ST	1982-84	5,963	R1100R, R850R	1994-2000	53,685
R80RT	1982-84	7,315	R1100RT	1995-2001	53,092
K100	1983-90	12,871	F650ST	1996-2000	13,349
K100RS	1983-92	47,470	R850C Classic/		
K100RT	1984-89	22,335	Avantgarde	1997-2001	1,505
R80	1984-95	13,815			

APPENDIX 2
SPECIFICATIONS OF PRODUCTION BMWs 1923–2003

Model	Years	Engine	Bore (mm)	Stroke (mm)	Displ (cc)	Comp ratio	Carburettor	HP
R32	1923-26	Twin Cyl SV	68	68	494	5.0:1	BMW 22mm	8.5 @ 3,200 rpm
R37	1925-26	Twin Cyl OHV	68	68	494	6.2:1	BMW 26mm	16 @ 4,000 rpm
R39	1925-27	Single Cyl OHV	68	68	247	6.0:1	BMW 20mm	6.5 @ 4,000 rpm
R42	1926-28	Twin Cyl SV	68	68	494	4.9:1	BMW 22mm	12 @ 3,400 rpm
R47	1927-28	Twin Cyl OHV	68	68	494	5.8:1	BMW 22mm	18 @ 4,000 rpm
R52	1928-29	Twin Cyl SV	63	78	486	5.0:1	BMW 22mm	12 @ 3,400 rpm
R62	1928-29	Twin Cyl SV	78	78	745	5.5:1	BMW 22mm	18 @ 3,400 rpm
R57	1928-30	Twin Cyl OHV	68	68	494	5.8:1	BMW 24mm	18 @ 4,000 rpm
R63	1928-29	Twin Cyl OHV	83	68	736	6.2:1	BMW 24mm	24 @ 4,000 rpm
R11	1929-34	Twin Cyl SV	78	78	736	5.5:1	BMW 24mm Sum 24mm 2 Amal 25mm	18 @ 3,400 rpm (20 @ 4,000 rpm)
R16	1929-34	Twin Cyl OHV	83	68	745	6.5:1 (7:1)	BMW 26mm 2 Amal 25mm	25 @ 4,000 rpm (33 @ 4,000 rpm)
R2	1931-36	Single Cyl OHV	63	64	198	6.7:1	Sum 19mm Amal 18.2mm	6 @ 3,500 rpm (8@ 4,500 rpm)
R4	1932-37	Single Cyl OHV	78	84	398	5.7:1	Sum 25mm	12 @ 4,000 rpm (14@ 4,000 rpm)
R3	1936	Single Cyl OHV	68	84	305	6.0:1	Sum 25mm	11 @ 4,200 rpm
R12	1935-42	Twin Cyl SV	78	78	745	5.2:1	Sum 25mm (2 Amal 24mm)	18 @ 3,400 rpm (20 @ 4,000 rpm)
R17	1935-37	Twin Cyl OHV	83	68	736	6.5:1	2 Amal 25.4mm	33 @ 5,000 rpm
R5	1936-37	Twin Cyl OHV	68	68	494	6.7:1	2 Amal 22.2mm	24 @5,800 rpm
R6	1937	Twin Cyl SV	70	78	596	6:1	2 Amal 22.2mm	18 @ 4,500 rpm
R35	1937-40	Single Cyl OHV	72	84	342	6.0:1	Sum 22mm	14 @4,500 rpm
R20	1937-38	Single Cyl OHV	60	68	192	6:1	Amal 18.2mm	8@5,400 rpm
R23	1938-40	Single Cyl OHV	68	68	247	6:1	Amal 18.2mm	10@5,400 rpm
R51	1938-40	Twin Cyl OHV	68	68	494	6.7:1	2 Amal 22.2mm	24 @ 5,800 rpm
R66	1938-41	Twin Cyl OHV	69.8	78	597	6.8:1	2 Amal 23.8mm	30 @5,300 rpm
R61	1938-41	Twin Cyl SV	70	78	596	5.7:1	2 Amal 22.2mm	18 @4,800 rpm
R71	1938-41	Twin Cyl SV	78	78	745	5.5:1	2 Graetzin 24mm	22 @4,600 rpm
R75	1941-44	Twin Cyl OHV	78	78	745	5.8:1	2 Graetzin 24mm	26 @4,000 rpm
R24	1948-50	Single Cyl OHV	68	68	247	6.75:1	Bing 22mm	12@5,600 rpm
R25	1950-51	Single Cyl OHV	68	68	247	6.5:1	Bing 24mm	12@5,600 rpm
R25/2	1951-53	Single Cyl OHV	68	68	247	6.5:1	Bing 22mm or SAWE 22mm	12@5,800 rpm
R25/3	1953-56	Single Cyl OHV	68	68	247	6.5:1	Bing 24mm or SAWE 24mm	12@5,600 rpm
R51/2	1950-51	Twin Cyl OHV	68	68	494	6.4:1	2 Bing 22mm	24 @5,800 rpm
R51/3	1951-54	Twin Cyl OHV	68	68	494	6.3:1	2 Bing 22mm	24 @5,800 rpm
R67	1951	Twin Cyl OHV	72	73	594	5.6:1	2 Bing 24mm	26 @5,500 rpm
R67/2	1952-54	Twin Cyl OHV	72	73	594	6.5:1	2 Bing 24mm	28 @5,600 rpm

Trans	Front Suspension	Rear Suspension	Front tyre	Rear tyre (mm)	Wheelbase (kg)	Weight	Model
3-speed	Cantilever Spring	Rigid	26x3	26x3	1380	122	R32
3-speed	Cantilever Spring	Rigid	26x3	26x3	1380	134	R37
3-speed	Plate Spring	Rigid	27x3.5	27x3.5	1232	110	R39
3-speed	Plate Spring	Rigid	26x3.5	26x3.5	1410	126	R42
3-speed	Plate Spring	Rigid	26x3	26x3	1410	130	R47
3-speed	Plate Spring	Rigid	26x3.5	26x3.5	1400	152	R52
3-speed	Plate Spring	Rigid	26x3.5	26x3.5	1400	155	R62
3-speed	Plate Spring	Rigid	26x3.5	26x3.5	1400	150	R57
3-speed	Plate Spring	Rigid	26x3.5	26x3.5	1400	152	R63
3-speed	Plate Spring	Rigid	26x3.5	26x3.5	1380	162	R11
3-speed	Plate Spring	Rigid	26x3.5	26x3.5	1380	165	R16
3-speed	Cantilever Spring	Rigid	25x3	25x3	1320 (1303)	130	R2
3-speed (4-speed)	Plate Spring	Rigid	26x3.5	26x3.5	1300	137	R4
4-speed	Cantilever Spring	Rigid	26x3.5	26x3.5	1320	149	R3
4-speed	Telescopic Fork	Rigid	3.5x19	3.5x19	1380	185	R12
4-speed	Telescopic Fork	Rigid	3.5x19	3.5x19	1380	183	R17
4-speed	Telescopic Fork	Rigid	3.5x19	3.5x19	1400	165	R5
4-speed	Telescopic Fork	Rigid	3.5x19	3.5x19	1400	175	R6
4-speed	Telescopic Fork	Rigid	3.5x19	3.5x19	1400	155	R35
3-speed	Telescopic Fork	Rigid	3x19	3x19	1330	130	R20
3-speed	Telescopic Fork	Rigid	3x19	3x19	1330	135	R23
4-speed	Telescopic Fork	Plunger	3.5x19	3.5x19	1400	182	R51
4-speed	Telescopic Fork	Plunger	3.5x19	3.5x19	1400	187	R66
4-speed	Telescopic Fork	Plunger	3.5x19	3.5x19	1400	184	R61
4-speed	Telescopic Fork	Plunger	3.5x19	3.5x19	1400	187	R71
4-speed + Rev	Telescopic Fork	Rigid	4.5x16	4.5x16	1444	400 With SC	R75
4-speed	Telescopic Fork	Rigid	3x19	3x19		130	R24
4-speed	Telescopic Fork	Plunger	3.25x19	3.25x19		140	R25
4-speed	Telescopic Fork	Plunger	3.25x19	3.25x19	1335	142	R25/2
4-speed	Telescopic Fork	Plunger	3.25x18	3.25x18	1365	150	R25/3
4-speed	Telescopic Fork	Plunger	3.5x19	3.5x19		185	R51/2
4-speed	Telescopic Fork	Plunger	3.5x19	3.5x19	1400	190	R51/3
4-speed	Telescopic Fork	Plunger	3.5x19	3.5x19	1400	192	R67
4-speed	Telescopic Fork	Plunger	3.5x19	3.5x19	1400	192	R67/2

Model	Years	Engine	Bore (mm)	Stroke (mm)	Displ (cc)	Comp ratio	Carburettor	HP
R67/3	1955-56	Twin CylOHV	72	73	594	6.5:1	2 Bing 24mm	28 @5,600 rpm
R68	1952-54	Twin Cyl OHV	72	73	594	8:1	2 Bing 26mm	35 @7,000 rpm
R50	1955-60	Twin Cyl OHV	68	68	494	6.8:1	2 Bing 24mm	26 @5,800 rpm
R69	1955-60	Twin Cyl OHV	72	73	594	8:1	2 Bing 26mm	35 @7,000 rpm
R60	1956-60	Twin Cyl OHV	72	73	594	6.5:1	2 Bing 24mm	28 @5,600 rpm
R26	1956-60	Single Cyl OHV	68	68	247	7.5:1	Bing 26mm	15@6,400 rpm
R50/2	1960-69	Twin Cyl OHV	68	68	494	7.5:1	2 Bing 24mm	26 @5,800 rpm
R60/2	1960-69	Twin Cyl OHV	72	73	594	7.5:1	2 Bing 24mm	30 @5,800 rpm
R27	1960-66	Single Cyl OHV	68	68	247	8.2:1	Bing 26mm	18@7,400 rpm
R50S	1960-62	Twin Cyl OHV	68	68	494	9.5:1	2 Bing 26mm	35 @7,650 rpm
R69S	1960-69	Twin Cyl OHV	72	73	594	9.5:1	2 Bing 26mm	42 @7,000 rpm
R50US	1967-69	Twin Cyl OHV	68	68	494	7.5:1	2 Bing 24mm	26 @5,800 rpm
R60US	1967-69	Twin Cyl OHV	72	73	594	7.5:1	2 Bing 24mm	30 @5,800 rpm
R69US	1967-69	Twin Cyl OHV	72	73	594	9.5:1	2 Bing 26mm	42 @7,000 rpm
R50/5	1969-73	Twin Cyl OHV	67	70.6	498	8.6:1	2 Bing 26mm	32 @6,400 rpm
R60/5	1969-73	Twin Cyl OHV	73.5	70.6	599	9.2:1	2 Bing 26mm	40 @6,400 rpm
R75/5	1969-73	Twin Cyl OHV	82	70.6	745	9:1	2 Bing CV 32mm	50 @6,200 rpm
R90S	1973-76	Twin Cyl OHV	90	70.6	898	9.5:1	2 Dell'Orto 38mm	67 @7,000 rpm
R90/6	1973-76	Twin Cyl OHV	90	70.6	898	9:1	2 Bing CV 32mm	60 @6,500 rpm
R75/6	1973-76	Twin Cyl OHV	82	70.6	745	9:1	2 Bing CV 32mm	50 @6,200 rpm
R60/6	1973-76	Twin Cyl OHV	73.5	70.6	599	9.2:1	2 Bing 26mm	40 @6,400 rpm
R100RS	1976-84	Twin Cyl OHV	94	70.6	980	9.5:1	2 Bing CV 40mm	70 @7,250 rpm
R100S	1976-78	Twin Cyl OHV	94	70.6	980	9.5:1	2 Bing CV 40mm	65 (70) @6,600 (7250) rpm
R100/7	1976-78	Twin Cyl OHV	94	70.6	980	9:1	2 Bing CV 32mm	60 @6,500 rpm
R75/7	1976-77	Twin Cyl OHV	82	70.6	745	9:1	2 Bing CV 32mm	50 @6,200 rpm
R60/7	1976-78	Twin Cyl OHV	73.5	70.6	599	9.2:1	2 Bing 26mm	40 @6,400 rpm
R80/7	1977-84	Twin Cyl OHV	84.8	70.6	797	9.2:1 (8:1)	2 Bing CV 32mm	55 (50) @7,000 (7250) rpm
R100RT	1978-84	Twin Cyl OHV	94	70.6	980	9.5:1	2 Bing CV 40mm	70 @7,250 rpm
R100T	1978-80	Twin Cyl OHV	94	70.6	980	9.5:1	2 Bing CV 40mm	65 @6,600 rpm
R45	1978-85	Twin Cyl OHV	70	61.5	473	9.2:1 (8.2:1)	2 Bing CV 28mm (26mm)	35 (27) @7,250 (6500) rpm
R65	1978-85	Twin Cyl OHV	82	61.5	649	9.2:1	2 Bing CV 32mm	45 (50) @7,250 rpm
R100CS	1980-84	Twin Cyl OHV	94	70.6	980	9.5:1	2 Bing CV 40mm	70 @7,250 rpm
R100	1980-84	Twin Cyl OHV	94	70.6	980	8.2:1	2 Bing CV 40mm	67 @7,000 rpm
R80G/S	1980-87	Twin Cyl OHV	84.8	70.6	797	8.2:1	2 Bing CV 32mm	50 @6,500 rpm
R65LS	1981-85	Twin Cyl OHV	82	61.5	649	9.2:1	2 Bing CV 32mm	50 @7,250 rpm
R80ST	1982-84	Twin Cyl OHV	84.8	70.6	797	8.2:1	2 Bing CV 32mm	50 @6,500 rpm
R80RT	1982-84	Twin Cyl OHV	84.8	70.6	797	8.2:1	2 Bing CV 32mm	50 @6,500 rpm
K100	1983-90	Four Cyl DOHC	67	70	987	10.2:1	Bosch LE Injection	90 @8,000 rpm
K100RS	1983-92	Four Cyl DOHC	67	70	987	10.2:1	Bosch LE Injection	90 @8,000 rpm
K100RT	1984-89	Four Cyl DOHC	67	70	987	10.2:1	Bosch LE Injection	90 @8,000 rpm
R80	1984-95	Twin Cyl OHV	84.8	70.6	797	8.2:1	2 Bing CV 32mm	50 @6,500 rpm

Trans	Front Suspension	Rear Suspension	Front tyre	Rear tyre (mm)	Wheelbase (kg)	Weight	Model
4-speed	Telescopic Fork	Plunger	3.5x19	3.5x19 4.00x18	1400	192	R67/3
4-speed	Telescopic Fork	Plunger	3.5x19	3.5x19	1400	190	R68
4-speed	Swingarm	Swingarm	3.5x18	3.5x18	1415	195	R50
4-speed	Swingarm	Swingarm	3.5x18	3.5x18	1415	202	R69
4-speed	Swingarm	Swingarm	3.5x18	3.5x18	1415	195	R60
4-speed	Swingarm	Swingarm	3.25x18	3.25x18	1390	158	R26
4-speed	Swingarm	Swingarm	3.5x18	3.5x18	1415	195	R50/2
4-speed	Swingarm	Swingarm	3.5x18	3.5x18	1415	195	R60/2
4-speed	Swingarm	Swingarm	3.25x18	3.25x18	1390	162	R27
4-speed	Swingarm	Swingarm	3.5x18	4.0x18	1415	198	R50S
4-speed	Swingarm	Swingarm	3.5x18	4.0x18	1415	202	R69S
4-speed	Telescopic Fork	Swingarm	3.5x18	4.0x18	1427	195	R50US
4-speed	Telescopic Fork	Swingarm	3.5x18	4.0x18	1427	195	R60US
4-speed	Telescopic Fork	Swingarm	3.5x18	4.0x18	1427	199	R69US
4-speed	Telescopic Fork	Swingarm	3.25x19	4.0x18	1385 (1435)	205 (200)	R50/5
4-speed	Telescopic Fork	Swingarm	3.25x19	4.0x18	1385 (1435)	210 (205)	R60/5
4-speed	Telescopic Fork	Swingarm	3.25x19	4.0x18	1385 (1435)	210 (205)	R75/5
5 speed	Telescopic Fork	Swingarm	3.25x19	4.0x18	1465	215	R90S
5 speed	Telescopic Fork	Swingarm	3.25x19	4.0x18	1465	210	R90/6
5 speed	Telescopic Fork	Swingarm	3.25x19	4.0x18	1465	210	R75/6
5 speed	Telescopic Fork	Swingarm	3.25x19	4.0x18	1465	210	R60/6
5 speed	Telescopic Fork	Swingarm	3.25x19	4.0x18	1465	230	R100RS
5 speed	Telescopic Fork	Swingarm	3.25x19	4.0x18	1465	220	R100S
5 speed	Telescopic Fork	Swingarm	3.25x19	4.0x18	1465	215	R100/7
5 speed	Telescopic Fork	Swingarm	3.25x19	4.0x18	1465	215	R75/7
5 speed	Telescopic Fork	Swingarm	3.25x19	4.0x18	1465	215	R60/7
5 speed	Telescopic Fork	Swingarm	3.25x19	4.0x18	1465	215	R80/7
5 speed	Telescopic Fork	Swingarm	3.25x19	4.0x18	1465	234	R100RT
5 speed	Telescopic Fork	Swingarm	3.25x19	4.0x18	1465	215	R100T
5 speed	Telescopic Fork	Swingarm	3.25x18	4.0x18	1400	205	R45
5 speed	Telescopic Fork	Swingarm	3.25x18	4.0x18	1400	205	R65
5 speed	Telescopic Fork	Swingarm	3.25x19	4.0x18	1465	220	R100CS
5 speed	Telescopic Fork	Swingarm	3.25x19	4.0x18	1465	218	R100
5 speed	Telescopic Fork	Monolever Swingarm	3.00x21	4.0x18	1465	186 (205)	R80G/S
5 speed	Telescopic Fork	Swingarm	3.25x18	4.0x18	1400	207	R65LS
5 speed	Telescopic Fork	MonoleverSwingarm	100/90x19	4.0x18	1446	198	R80ST
5 speed	Telescopic Fork	Swingarm	3.25x19	4.0x18	1465	235	R80RT
5 speed	Telescopic Fork	MonoleverSwingarm	100/90x18	130/90x17	1516	239	K100
5 speed	Telescopic Fork	MonoleverSwingarm	100/90x18	130/90x17	1516	249	K100RS
5 speed	Telescopic Fork	MonoleverSwingarm	100/90x18	130/90x17	1516	253	K100RT
5 speed	Telescopic Fork	MonoleverSwingarm	90/90x18	120/90x18	1447	210	R80

Model	Years	Engine	Bore (mm)	Stroke (mm)	Displ (cc)	Comp ratio	Carburettor	HP
R80RT	1984-95	Twin Cyl OHV	84.8	70.6	797	8.2:1	2 Bing CV 32mm	50 @ 6,500 rpm
R65	1985-93	Twin Cyl OHV	82	61.5	649	8.7:1 (8.4:1)	2 Bing CV 32mm (26mm)	48 (27) @7,250 rpm
K75C	1985-90	Three Cyl DOHC	67	70	740	11:1	Bosch LE Injection	75 @ 8,500 rpm
K75S	1985-95	Three Cyl DOHC	67	70	740	11:1	Bosch LE Injection	75 @8,500 rpm
K75	1986-96	Three Cyl DOHC	67	70	740	11:1	Bosch LE Injection	75 @8,500 rpm
K100LT	1986-91	Four Cyl DOHC	67	70	987	10.2:1	Bosch LE Injection	90 @8,000 rpm
R100RS	1986-92	Twin Cyl OHV	94	70.6	980	8.45:1	2 Bing CV 32mm	60 @6,500 rpm
R100GS	1987-96	Twin Cyl OHV	94	70.6	980	8.5:1	2 Bing CV 40mm	60 @6,500 rpm
R80GS	1987-96	Twin Cyl OHV	84.8	70.6	797	8.2:1	2 Bing CV 32mm	50 @6,500 rpm
R65GS	1987-92	Twin Cyl OHV	82	61.5	649	8.4:1	2 Bing CV 26mm	27 @5,500 rpm
R100RT	1987-96	Twin Cyl OHV	94	70.6	980	8.45:1	2 Bing CV 32mm	60 @6,500 rpm
K1	1989-93	Four Cyl DOHC 4v	67	70	987	11:1	Bosch Motronic Injection	100 @8,000 rpm
K100RS	1989-92	Four Cyl DOHC 4v	67	70	987	11:1	Bosch Motronic Injection	100 @8,000 rpm
K75RT	1989-96	Three Cyl DOHC	67	70	740	11:1	Bosch LE Injection	75 @8,500 rpm
R100R (Mystic)	1991-96	Twin Cyl OHV	94	70.6	980	8.5:1	2 Bing CV 40mm	60 @6,500 rpm
R80R	1992-94	Twin Cyl OHV	84.8	70.6	797	8.2:1	2 Bing CV 32mm	50 (34,27) @6,500 rpm
K1100LT	1991-98	Four Cyl DOHC 4v	70.5	70	1092	11:1	Bosch Motronic Injection	100 @7,500 rpm
K1100RS	1992-96	Four Cyl DOHC 4v	70.5	70	1092	11:1	Bosch Motronic Injection	100 @7,500 rpm
R1100RS	1993-2001	Twin Cyl OHV 4v	99	70.5	1085	10.7:1	Bosch Motronic Injection	90 @7,250 rpm
F650	1993-99	Single Cyl DOHV 4v	100	83	652	9.7:1	2 Mikuni 33mm	48 (34) @6,500 rpm
R1100GS	1993-99	Twin Cyl OHV 4v	99	70.5	1085	10.3:1	Bosch Motronic Injection	80 @6,750 rpm
R1100R	1994-2000	Twin Cyl OHV 4v	99	70.5	1085	10.3:1	Bosch Motronic Injection	80 @6,750 rpm
R850R	1994-2002	Twin Cyl OHV 4v	87.5	70.5	848	10.3:1	Bosch Motronic Injection	70 (34) @7,000 rpm
R1100RT	1995-2001	Twin Cyl OHV 4v	99	70.5	1085	10.7:1	Bosch Motronic Injection	90 @7,250 rpm
F650ST	1996-99	Single Cyl DOHV 4v	100	83	652	9.7:1	2 Mikuni 33mm	48 (34) @6,500 rpm
K1200RS	1997-	Four Cyl DOHC 4v	70.5	75	1171	11:5	Bosch Motronic Injection	130 (98)@8,750 rpm
R1200C	1997-	Twin Cyl OHV 4v	101	73	1170	10.0:1	Bosch Motronic Injection	61 @5,000 rpm
R1100S (Boxer Cup)	1998-(2003)	Twin Cyl OHV 4v	99	70.5	1085	11.3:1	Bosch MotronicInjection	98 @7,500 rpm
R850GS	1998-	Twin Cyl OHV 4v	87.5	70.5	848	10.3:1	Bosch Motronic Injection	70 (34) @7,000 rpm
K1200LT	1999-	Four Cyl DOHC 4v	70.5	75	1171	10:8	Bosch Motronic Injection	98 @6,750 rpm
R1150GS (Adventure)	1999-	Twin Cyl OHV 4v	101	70.5	1130	10.3:1	Bosch Motronic Injection	85 @6,750 rpm
R850C	1999-2002	Twin Cyl OHV 4v	87.5	70.5	848	10.3:1	Bosch Motronic Injection	50 (34) @7,000 rpm
F650GS (Dakar)	2000-	Single Cyl DOHV 4v	100	83	652	11.5:1	BMS	50 (34) @ 6,500 rpm
R1150R	2000-	Twin Cyl OHV 4v	101	70.5	1130	10.3:1	Bosch Motronic Injection	85 @6,750 rpm
C1	2000-	Single Cyl DOHV 4v	56.4	50	125	13:1	BMS	15@9,250
R1150RT	2001-	Twin Cyl OHV 4v	101	70.5	1130	11.3:1	Bosch Motronic Injection	95 @7,250 rpm
C1 200	2001-	Single Cyl DOHV 4v	62	58.8	176	11.5:1	BMS	18@9,000
R1150RS	2002-	Twin Cyl OHV 4v	101	70.5	1130	11.3:1	Bosch Motronic Injection	95 @7,250 rpm
F650CS	2002-	Single Cyl DOHV 4v	100	83	652	11.5:1	BMS	50 (34) @6,800 rpm
K1200GT	2003-	Four Cyl DOHC 4v	70.5	75	1171	11:5	Bosch Motronic Injection	130 (98) @8,750 rpm
R1200CL	2003-	Twin Cyl OHV 4v	101	73	1170	10.0:1	Bosch Motronic Injection	61 @5,000 rpm
R850R	2003-	Twin Cyl OHV 4v	87.5	70.5	848	10.3:1	Bosch Motronic Injection	70 (34) @7,000 rpm

Trans	Front Suspension	Rear Suspension	Front tyre	Rear tyre (mm)	Wheelbase (kg)	Weight	Model
5 speed	Telescopic Fork	MonoleverSwingarm	90/90x18	120/90x18	1447	227	R80RT
5 speed	Telescopic Fork	MonoleverSwingarm	90/90x18	120/90x18	1447	205	R65
5 speed	Telescopic Fork	MonoleverSwingarm	100/90x18	120/90x18	1516	227	K75C
5 speed	Telescopic Fork	MonoleverSwingarm	100/90x18	130/90x17	1516	229	K75S
5 speed	Telescopic Fork	MonoleverSwingarm	100/90x18	120/90x18	1516	227	K75
5 speed	Telescopic Fork	MonoleverSwingarm	100/90x18	130/90x17	1516	283	K100LT
5 speed	Telescopic Fork	MonoleverSwingarm	90/90x18	120/90x18	1447	229	R100RS
5 speed	Telescopic Fork	Paralever Swingarm	90/90x21	130/80x17	1514	210 (236)	R100GS
5 speed	Telescopic Fork	Paralever Swingarm	1.85x21	130/80x17	1514	210	R80GS
5 speed	Telescopic Fork	MonoleverSwingarm	1.85x21	4.0x18	1465	198	R65GS
5 speed	Telescopic Fork	MonoleverSwingarm	90/90x18	120/90x18	1447	229	R100RT
5 speed	Telescopic Fork	Paralever Swingarm	120/70x17	160/60x18	1565	234	K1
5 speed	Telescopic Fork	Paralever Swingarm	120/70x17	160/60x18	1565	235	K100RS
5 speed	Telescopic Fork	MonoleverSwingarm	100/90x18	130/90x17	1516	258	K75RT
5 speed	Telescopic Fork	Paralever Swingarm	110/80x18	140/80x17	1495	218 (215)	R100R
5 speed	Telescopic Fork	Paralever Swingarm	110/80x18	140/80x17	1495	217	R80R
5 speed	Telescopic Fork	Paralever Swingarm	110/80x18	140/80x17	1565	290	K1100LT
5 speed	Telescopic Fork	Paralever Swingarm	120/70x17	160/60x18	1565	268	K1100RS
5 speed	Telelever Fork	Paralever Swingarm	120/70x17	160/60x18	1473	239	R1100RS
5 speed	Telescopic Fork	Rising rate Swingarm	100/90x19	130/80x17	1480	189 (191)	F650
5 speed	Telelever Fork	Paralever Swingarm	110/80x19	150/70x17	1509	243	R1100GS
5 speed	Telelever Fork	Paralever Swingarm	120/70x17 110/80x18	160/60x17 150/70x17	1487	235	R1100R
5 speed	Telelever Fork	Paralever Swingarm	120/70x17 110/80x18	160/60x17 150/70x17	1487	235	R850R
5 speed	Telelever Fork	Paralever Swingarm	120/70x17	160/60x18	1485	282	R1100RT
5 speed	Telescopic Fork	Rising rate Swingarm	100/90x18	130/80x17	1465	191	F650ST
6 speed	Telelever Fork	Paralever Swingarm	120/70x17	170/60x17 180/55x17	1555	285	K1200RS
5 speed	Telelever Fork	Monolever Swingarm	110/90x18	170/80x15	1650	256	R1200C
6 speed	Telelever Fork	Paralever Swingarm	120/70x17	170/60x17 (180/55x17)	1478	229	R1100S
5 speed	Telelever Fork	Paralever Swingarm	110/80x19	150/70x17	1509	249	R850GS
5 speed	Telelever Fork	Paralever Swingarm	120/70x17	160/60x17	1633	378	K1200LT
6 speed	Telelever Fork	Paralever Swingarm	110/80x19	150/70x17	1509 (1501)	249 (253)	R1150GS
5 speed	Telelever Fork	Paralever Swingarm	110/90x18	170/80x15	1650	256	R850C
5 speed	Telescopic Fork	Rising rate Swingarm	100/90x19 (90/90x21)	130/80x17	1479 (1489)	193 (192)	F650GS
6 speed	Telelever Fork	Paralever Swingarm	120/70x17	170/60x17	1487	238	R1150R
Auto 2 stage	Telelever Fork	Swingarm	120/70x13	140/70x12	1488	185	C1
6 speed	Telelever Fork	Paralever Swingarm	120/70x17	170/60x17	1485	279	R1150RT
Auto 2 stage	Telelever Fork	Swingarm	120/70x13	140/70x12	1488	185	C1 200
6 speed	Telelever Fork	Paralever Swingarm	120/70x17	170/60x17	1473	248	R1150RS
5 speed	Telescopic Fork	Rising rate Monolever	110/70x17	160/60x17	1493	189	F650CS
6 speed	Telelever Fork	Paralever Swingarm	120/70x17	180/55x17	1555	300	K1200GT
6 speed	Telelever Fork	Monolever Swingarm	150/80x16	170/80x15	1641	308	R1200CL
6 speed	Telelever Fork	Paralever Swingarm	120/70x17	170/60x17	1487	238	R850R

INDEX

Photograph captions are in Italics.